THE FASCINATION OF SCIENCE & INVENTIONS

By Norman Renshaw

Lundarien Press

Published by Lundarien Press, UK
Copyright © Norman Renshaw 2015

ISBNs
978-1-910816-82-0 (paperback)
978-1-910816-79-0 (Kindle ebook)
978-1-910816-80-6 (Epub ebook)
978-1-910816-82-0 (PDF ebook)

Contents

Preface

This book attempts to pay tribute to the achievements of many gifted people whose theories, discoveries and inventions, sometimes reaching sheer genius have shaped the world. It is arranged mostly in chronological form to make it possible to trace the tremendous progress made in just a few centuries. Occasionally a name will be repeated as one scientific discipline finds an affinity with another and this makes it even more intriguing.

The high level of interest and knowledge among members of the public is displayed in all types of subjects in quiz sessions in clubs, pubs and on television etc. To cater for this a section of the book has been devoted to adding set questions, over 300, on the subjects covered in the book but many times that number may easily be derived from the narrative.

1. THE RAILWAY - PUTTING THE WORLD ON TRACK

1771-1833: Trevithick, Richard - English

Father of the Locomotive, Trevithick rejected conventional condensing engines and developed the principle of using high-pressure steam for locomotion. In 1801, using this system, he produced the world's first steam locomotive, his famous "Puffing Devil", to carry passengers by steam power (albeit not on rails). He is recorded as the first to introduce four other inventions - (a) the coupling of all wheels (b) introducing the steam jet in the chimney. (c) the first to make the return flue boiler and (d) the first to prove that the adhesion between wheels and rails provided sufficient traction on ordinary gradients. In 1804, he installed the world's first locomotive in Merthyr Tydfil, Wales, for hauling coal and steel.

1783-1831: Blenkinsop, John - English

In 1811, Blenkinsop patented a rack railway with a cogged driving wheel, which engaged with teeth cast on one side of the rails, enabling an engine of equal weight to Trevithick's to haul a load five times greater, particularly important for hauling heavy loads up steep gradients.

1779-1843: Hedley, William - English

Hedley took out a patent in 1813 for a locomotive which he named "Puffing Billy". Proved to be the most commercially viable to employ, it utilised the principle of adhesion between wheels and rails to provide traction (now accepted by all modern railways). It was in regular use at Wylam colliery, hauling coal to the docks at Leminton-on-Tyne. It was later installed in the South Kensington Museum.

1781-1848: Stephenson, George - English

It was George Stephenson who gave the world the railway system. His vision extended far beyond simply transporting coal from the pits and he designed a locomotive accordingly. Successful trials of his first locomotive were held in July 1814, when it drew a load weighing 30 tons at a speed of 4 miles per hour. Further improvements to the locomotive were made by his patent in 1815 and this eventually convinced doubtful investors in his plan to build a railway to link towns. In 1821, he was commissioned to build the Stockton & Darlington Railway which opened, first of all for freight in September 1825. This was the first public railway, an invention which English engineers helped to spread rapidly throughout the world, opening up countries and even continents. An outstanding achievement for someone who started his working life as a cowherd! In May of 1825, the Canterbury to Whitstable line was opened, for which Stephenson supplied the locomotive. Designed for both passengers and freight, it is claimed that it qualifies as the first passenger railway line in the world. Later that year, Stephenson and his son designed and built the Liverpool to Manchester railway, on which he installed his famous locomotive, "The Rocket". An early system of ticketing and timetables was evolved, obviously limited at the time, but it was the beginning of rail travel.

1803 - 1859: Stephenson, Robert - English

Robert Stephenson made improvements to the Rocket, built the first London to Birmingham railway and the Menai tubular rail bridge to Anglesey, as well as the railway bridge at Berwick-on Tweed and the high level bridge at Newcastle-upon-Tyne. His greatest achievement was to design and build the mighty Victoria bridge over the St. Lawrence river in Canada.

1806 - 1859: Brunel, Isambard K - English

Brunel was an on outstanding engineer in almost every way and, aged 20, he was resident engineer at the building of the first Thames Tunnel. In 1833, he was appointed to build the Great Western Railway, opening the first stage in 1838 and reaching Bristol, the home of those who financed it, in 1841. During this project he designed the Maidenhead Bridge and the famous, and controversial, Box Tunnel. He was responsible for over 1,000 miles of railway built in the West Country, South Wales and Ireland. He built railway lines in Italy and was advisor on construction of the railways in East Bengal and Australia.

1814 July 25th. G. Stephenson tested his first successful rail locomotive.

1815 G. Stephenson patented his improved engine.

1821 Appointed Chief engineer to Stockton & Darlington railway

1823 Opened his engine works in Newcastle

1825 Stockton & Darlington railway (the world's first public railway) opened in September.

1830 World's first passenger railway station opened at Liverpool Rd., Manchester.

1833 First bogie locomotive was sold to USA

1837 The prototype telegraph of Wheatstone/Cook was applied to railways

1837 Morse code used to apprise Queen Victoria's relatives of her succession

1838 Journey time London-Manchester =11 hours.

1838 First railway excursion organised by Thomas Cook (from Leicester)

1839 George Bradshaw produced his famous railway time table and travel guide

1841 Hall's brick arch fire-box for smokeless combustion introduced

1841 Semaphore arm rail signal introduced on London/Croydon line

1842 Thomas Cook publicised his regular railway excursions.

1845 Workable elec-telegraph system by Wheatstone & Cook adopted by railways

1844 Samuel Morse fitted his telegraph and code on Baltimore / Washington line

1860 Water scoop introduced, permitting water refilling of tender while travelling

1862 Trials began on Bessemer steel rail

1864 Testing on Kendall compressed air-brake began

1868 By this date there were more than 20,000 miles of railway track in Britain

1869 Westinghouse patented the straight-air brake

1871 Mount Cenis railway tunnel France to Italy opened (approx.8.5 miles long)

1873 Sleeping cars introduced between England and Scotland.

1875 Oil & Gas lighting on Metropolitan line (invented by G. Pentch)

1879 Dining cars introduced in GB

1883 First corridor coaches incorporated

1883 Workmen's tickets were introduced

1895 Diesel designed the renowned Diesel engine.

1938 Highest speed British steam locomotive "Mallard" reached 126mph while hauling a load of 240 tons

Over a period there were other entrants to the transport industry, including the Omnibus in 1622, when Louis 1V granted a patent to Blaise Pascal to run an omnibus service in Paris, which quickly failed. Another, which begun in 1829, was an omnibus service in London by Shillibeer, who also introduced a top deck, improved by Miller in 1857, who enclosed the upper deck and also improved access by designing a winding stairway. In the mid 19th. century, powered by electricity, the tramcar arrived. It carried people within towns and urban districts cheaply and efficiently. Congestion resulted by the increasing numbers of motor cars and in the 1950's trams were replaced by buses. Birkenhead claims to have been the first to install trams in Britain in 1860.

2. THE AUTOMOBILE – A PERSONAL STAGECOACH

The first vehicles intended for road transport were based on the developments in the use of steam, Gurney and Murdock in England and Cugnot in France were building steam-driven road vehicles but they were overtaken by the search for an engine designed to use different fuels to produce greater efficiency. The internal combustion engine was the aim, with the prospect of the motor car being the result. When the motor car did arrive it excited almost everyone except the British government of the time, who were strangely hostile to its introduction into England from the very beginning. It was possibly influenced by those dominating the transport industry or perhaps imagining that they themselves might lose the tax receipts from that industry they restricted its speed by law to 4 miles per hour and insisted that a man with a red flag was required to walk ahead of the car at all times (see Simms.)

1725 - 1804: Cugnot, Joseph - French

Cugnot, a soldier-cum-inventor, in 1769 after experimenting with the addition of a steam engine to propel gun carriages he built the first high- pressure steam-driven road vehicle. It was designed like a tricycle and the engine design owed much to his French compatriot Papin but he had embarked on his tests without first incorporating a safety valve to control the pressure consequently his first efforts literally went up in a cloud of steam. Nevertheless he finally succeeded in building a road travelling steam engine.

1800 - 1860: Goodyear, Charles - American

Goodyear discovered the method of hot vulcanisation for rubber and it was first patented in 1844 but legal difficulties prevented his American patents being ratified until 1852. It coincided with the growth in the automotive industry and eventually enabled pneumatic tyres for all manner of vehicles to be manufactured. He spent much time fruitlessly trying to enforce his patents.

1815 - 1893: Beau de Rochas, Alphonse - French

In 1862, Rochas is claimed to be the first to issue a patent of the theory of the four-stroke internal combustion engine. He apparently took no further interest in the project until the success and practicality of Otto's engine was proved so conclusively.

1822 - 1900: Lenoir, Etienne - Belgian

Lenoir is reported as being the first to build an internal combustion engine. It was a two stroke and low powered, employing coal gas as a fuel initially though it is probable that he did eventually manage to convert it to using a liquid fuel.

1832 - 1891: Otto, Nikolaus - German

After a number of years of development and investment, in 1867, Otto produced the first practical internal combustion engine. It was a gas fuelled four stroke engine which it attracted many buyers. The French, however, produced documents dated 1862, showing that a patent based on a design of similar principle to that used by Otto had been registered in France in the name of Beau de Rochas. In spite of the fact that during the years that the plans of de Rochas had lain dormant they had never been subjected to test or manufacture, Otto's patent was rescinded.

1834 - 1900: Daimler, Gottlieb - German

Daimler, a talented engineer, in 1882, began his own company to build engines to his own design. It incorporated his invention of the carburettor which was used in conjunction with a petroleum based fuel. These changes circumvented any patent objections and he patented his engine in 1885. He used his new engine to power a sequence of new inventions, the motor-cycle, the motorboat and ultimately, a motor car.

1858 - 1913: Diesel, Rudolph - German

Diesel's family were living in France and being German were expelled from France in 1870, at the start of the Franco-Prussian war. They came to England where Rudolph received his education after which he returned to Germany. In 1892, he designed and patented an engine which used a crude form of oil as a combustible. It developed more power than steam engines therefore suitable for ship propulsion. He failed to interest the German government in its use for Naval vessels but was invited by the British Admiralty in 1913 to visit London for discussions. For this purpose he boarded a ship from Germany but was nowhere to be found when the ship arrived in England and was presumed to have been lost overboard.

1861 - 1942: Bosch, Frederick R - German

Attracted by the reputation of Thomas Edison he went to the U.S.A to work and train, before returning to Germany to manufacture all manner of electrical and mechanical devices ancillary to the improvement of the motor vehicle. Among them the magneto and the spark plug. The former reputedly in cooperation with Simms and the latter produced by a member of his company G. Honold in 1902.

1863 - 1944: Simms, Frederick - English

Against opposition from the British government of the time, Simms introduced the internal combustion engine into Britain. In 1891, having been refused permission to demonstrate it on land, he hired a Thames launch on which to make his presentation. In 1896, he manufactured "The Simms 2.5 Horseless carriage" and it was during that year that the flagman was abolished and the law restricting the speed to 4 miles per hour was extended to 12 mph. It was facilitated by the passing of the "Light (Road) Locomotives Act 1896", which owed much to the campaigning by Simms as founder of the Society of Motor Manufacturers and Traders.

It was to celebrate these changes that in the November of that year the London-Brighton Run was inaugurated. As a consequence, he is regarded as the Father of the British motor industry. In addition, he worked in cooperation with Bosch, to invent magneto-ignition and his invention of the Simms coupling simplified the timing mechanism.

1863 - 1947: Ford, Henry - USA

Ford founded his company in 1903 and, within five years, his model T Ford was so successful that he introduced 24 hour mass-production to keep up with demand. One brilliant time and cost-saving inspiration of his was to instruct all suppliers of parts to pack them in wooden cases, all six sides of which had to be of exact dimensions. After the contents were unpacked each of the six sides was carefully separated to become a ready-made floor of a model-T Ford, a perfect example of lateral thinking.

1877 - 1910: Charles S. Rolls - English
1863 - 1933: Frederick H. Royce - English

Setting up in 1904, with Rolls providing the finance and Royce the engineering, they built the incomparable Rolls-Royce motor car. Royce aspired to perfection and devoted his skill to this end. Rolls admired this approach and, together, they achieved their respective goals. The company now produces the famous RR.aero-engines. The car is still manufactured to the same high quality but is now owned by a German company.

1884 - 1929: Benz, Karl F - German

Benz patented a three-wheeled vehicle in 1886, which is considered to be the first practical automobile driven by an internal combustion engine. In 1893, he produced a four-wheeled version. The Daimler-Benz Company was formed in 1926.

3. PEDAL POWER – GOODBYE, SHANK'S PONY

After the development of the wheel the next tentative steps toward developing a means by which people could propel themselves along faster and, without expending as much energy as walking, were seen even on the walls of ancient sites. The modern efforts have continued and in 1779 a very crude construction called a velocipede was made in France. It was simply two wheels linked by a bar on which the rider sat and propelled himself by pushing his feet against the ground. It was no improvement on the images found drawn on those ancient walls but it did revive the endeavour. Later results such as those listed below followed.

Drais von Sauerbraun from Germany submitted a machine which differed only in that it had a swivelling front wheel, allowing it to be steered. This machine was known as the "Draisine". In England, it was called the Hobby Horse.

Several claimed the honour of being the first to devise a method of propulsion but the strongest claim is by Kirkpatrick MacMillan, a Scottish blacksmith who, in 1839, connected pedals to the front wheel.

They did not move in a circular motion but were linked by rods to the rear wheel and achieved traction by a shuttling motion.

Although the rubber pneumatic tyre was invented later, the first to take out a patent for a pneumatic tyre was Robert Wm. Thomson in 1845. He was making solid rubber tyres for London carriages and experimented with air filled leather tyres but, in spite of successful tests, his clients preferred the solid rubber type.

From 1865, the names of Pierre Michaux, his son Henri, and Pierre Lallement, produced the first machine to be classed as a bicycle. It was known in Britain as the "Penny Farthing", on account of the size of the front wheel compared to the rear and, in spite of its solid tyres, it became popular. The first cycling club (The Pickwick of London) was founded in 1870.

In 1871, an Englishman named John K. Starley built what was described as the first true bicycle - "The Ariel". It had pivotal steering mounted centrally and Stanley later invented tangentially-spoked wheels to add tensile strength to cope with stress.

Another English engineer, H.J. Lawson, is recorded as the first to build a chain-driven bicycle in 1874.

In 1888, a Scottish veterinarian, John B. Dunlop, while living in Ireland, had observed the Goodyear development of "hot vulcanisation", which was by then out of patent in Britain and he decided to use it to make rubber tyres which were an immediate success and it may be said – completed the cycle (no pun intended).

4. SPINNING AND WEAVING

Ancient crafts changed from being cottage industries to being a highly mechanised industry employing thousands, contrary to the fears of the Luddites, for which the first recorded inventor enjoys pride of place.

1550 - 1610: Lee, William - English

In about 1585, Lee devised a machine (aka a frame) for knitting stockings. The governing council of Elizabeth I. refused to grant him a licence to manufacture, so he took the invention to France, where Henry 1V was impressed and granted him manufacturing rights, which upset local stocking makers who looked unfavourably on the competition. Lee prospered until in 1610 King Henry was assassinated. By a strange co-incidence, Lee died in the same year as his Royal (but only) supporter.

1704 - 1764: Kay, John - English

Kay invented "The flying shuttle", a mechanical device for transferring the shuttle from one side of the cloth width to the other. His patent in 1733 not only quadrupled the speed of the loom it also needed only one operator. It laid the foundations of automatic weaving and much discontent as once again job losses were feared

1690 - 1759: Paul, Lewis - English
1700 - 1760: Wyatt, John - English

Lewis and John jointly invented the first powered spinning machine, patented 1738 and it was in general use for about 20 years, until being superseded by the Arkwright Frame. Paul also patented a carding machine in 1748, which automatically removed all foreign substances and laid the threads in one direction.

1709 - 1782: Vaucanson, Jacques - French

By using a punched-card system, Vaucanson invented a method of controlling and guiding the loom, but in practice with only limited success. The brilliance of the idea was recognised and adopted some years later by one of his countrymen who brought it to perfection, in automated form.

1720 - 1778: Hargreaves, James - English

Perhaps more an innovator than inventor of the "Spinning Jenny" in 1764, Hargreaves mounted the spinning wheels vertically and in multiples, enabling one operator to control a number of machines as opposed to just one. The mill owners were enthusiastic but workers were alarmed and destroyed much of his machinery, forcing him to move his business.

1732 - 1792: Arkwright, Richard - English

First a barber and later a wig salesman, Arkwright, who is generally regarded as the founder of the Lancashire cotton industry, studied the "Spinning Jenny" and, within four years, had re-designed it, embodying all his own ideas. The result in 1769 was "The Arkwright Frame", which was an outstanding success

1753 - 1827: Crompton, Samuel - English

In 1779, Crompton combined the attributes of Hargreave's Jenny and Arkwright's Frame and produced a machine which he named the "Mule." In addition, he introduced a spindle carriage, which eased the pressure on the thread, thus reducing the frequency of breakage. Brilliant as they were, they were still modifications and were easily plagiarised and he thus benefitted little from his endeavours. A sympathetic government acknowledged his contribution by an award of £5,000, which did not begin to reflect the value to mill owners who could now link machines together in their hundreds.

1735 - 1797: Strutt, Jedediah - English

Strutt became wealthy by improving upon the stocking frame made by Lee about 200 years earlier. His machine incorporated a mechanism which introduced ribbing, which could resist the embarrassing crumpling effect common in basic stocking design. He also built the first cotton mill in the Amber Valley, Derbyshire.

1743 - 1823: Cartwright, Edmund - English

Cartwright was the inventor of the Power Loom for weaving in 1785, which revolutionised productivity. How Cartwright, a country vicar, was able to devise such a major improvement is still a mystery. It caused hostility among factory workers, who broke into his premises in Manchester and burned 400 looms in a single night, which almost bankrupted him. The earnings this invention brought to both industry and country in international trade were massive. In recognition, a grateful government awarded him £10,000 for his services to industry.

1752 - 1834: Jacquard, Joseph M - French

Jacquard studied and expanded the idea of Vaucanson, which proposed a design that would enable a punched-card system to automatically control the replication of chosen designs without any involvement by an operator. He finally produced a machine of great ingenuity. By using sets of punched cards with each set of cards dedicated to a particular pattern, the loom could be instantly programmed. It could then replicate a particular pattern of even the most intricate form, such the famous "Paisley" pattern.

1783 - 1861: Heathcoat, John - English

Originally from Derby, and a blacksmith, Heathcoat moved to Nottingham, where he took a job with a lace-making company. Lace-making was an individual skill which needed extensive training time but his invention of a lace-making machine not only required little training but reduced the number of operators required. Fear of unemployment incited the workforce to destroy his factory. As a result he moved the company to Tiverton in Devon, where there was no previous tradition of lace manufacture.

1795 - 1860: Roberts, Richard - (Possibly Welsh)

A former miner, Roberts worked in London under the tuition of Maudsley, later moving to Manchester to work in the cotton industry, during which time he devised the " Self-Acting Mule", which added even further improvement to Crompton's original invention.

5. THE AGE OF STEAM

1601 - 1667: The Marquis of Worcester - English

In 1645, the Marquis of Worcester invented a steam pump capable of raising water from a depth of 40 feet. He also wrote a book envisaging a wide range of possibilities, which he entitled "A Century of Inventions" and completed it while being held as a political prisoner in the Tower of London by Cromwell's government.

1647 - 1712: Papin, Denis - French

An émigré from France, Papin is thought to have been the first to suggest the technique of using the piston and cylinder to produce steam pressure. He continued his experiments with steam, resulting in his invention of the pressure cooker in 1679.

1650 - 1715: Savery, Thomas - English

The first practical steam pump was invented and patented in 1698. It incorporated hand-operated valves, used condensed steam to create suction and was intended for use in mines, to clear the water caused through seepage.

1663 - 1729: Newcomen, Thomas - English

In 1715, Newcomen developed a much more efficient steam engine than his predecessors. His design was to compress the steam by means of a piston and this design was central to the improved machines which other developers such as James Watt produced.

1736 - 1819: Watt, James - Scottish

Trained in London as an instrument maker, Watt returned to Scotland and began trading in or repairing various scientific instruments, particularly for students.In 1759, he was approached by John Roebuck, an English industrialist, to develop a more powerful pump for his deep mines in Scotland. Watt was provided with a model of Newcomen's steam engine, to study, and for six years Roebuck arranged for parts to be made and supported him financially during his studies. Almost at the point of success, Roebuck suddenly became bankrupt and obviously no longer able to continue to provide finance but, in order that the project was not abandoned, he introduced Watt to Mathew Boulton a major industrialist in Birmingham, who agreed not only to continue to finance the project but also provide the engineering facilities to bring it to fruition. In 1768, the improved steam engine was completed, patented and full-size drawings completed. The final product was a great success and contributed much to the Industrial Revolution

1753 - 1815: Hornblower, Jonathon C - English

Hornblower invented the first reciprocating compound steam engine in 1781 and, incorporating two cylinders, it generated greater power than all others, including that of Watt. He was subjected to a legal challenge, for infringement of patent, by Boulton and Watt and even though he was convinced that their challenge was unsubstantiated lack of capital rendered him unable to defend the case and he was forced to concede.

1766 - 1837: Woolf, Arthur - English

Another product of the great Bramah workshop, Woolf, in 1804, invented and patented a compound high-pressure engine which developed a thermal output efficiency of approximately 7.5% , almost twice that of the Watt engine. However, having witnessed the demise of the Hornblower invention because of the costly legal wrangling over patent rights, he did not pursue it. In 1810, when the Boulton & Watt patent protection had lapsed, he then patented it successfully.

1771 - 1833: Trevithick, Richard - English

Having found that Watt's low pressure machine was too large and not powerful enough to reduce the water which accumulated in deep mines, Cornish miners still urgently sought a more powerful engine. Trevithick took the opposite view to Watt regarding the safety of developing a high pressure steam engine and, in 1800, he provided a high-pressure engine which generated the required power.

1793 - 1895: Gurney, Goldsworthy - English

Weighing almost 2 tons and at a speed which was said to be 15 miles per hour, In 1830 c. Gurney built a road vehicle powered by steam, which he drove from Bath to London and back. He increased the number of vehicles and started a passenger service as far as Gloucester but his enterprise ended when the Inland Revenue acknowledged his service as a vehicle but only for the purpose of imposing taxation.

1854 - 1931: Parsons, Charles - English

Parsons' invention of the Steam Turbine made it possible to build ships of greater size and tonnage. They were adopted by the Admiralty and also used in the passenger liners. They powered most of the great liners of the 20th century including the great liner the Queen Elizabeth which was built on the Clyde just prior to WW2.

6. HOW AVIATION GOT OFF THE GROUND!

As far back as 100.BC, an enterprising Greek named Daedalus, after observing the grace and manoeuvrability of sea-birds in flight around the coast of Crete, reasoned that, by copying the shape of the wings, gliding might be possible. In his Greek version of the garden shed, he set himself to make wings which could be strapped to his body but sadly, before he had time to test this theory, his young son, Icarus, discovered them and, with the presumption and impetuosity of youth, decided to be the world's first test pilot. Strapping them to his body he launched himself off a cliff and instead he became the world's first aerial casualty.

Probably one of the first man-made items was the spear which, if carefully shaped, would obey the laws of aero-dynamics to guarantee a successful result. Without doubt the Australian aborigines had already mastered aerodynamics by the invention of the Boomerang, hundreds or perhaps, thousands, of years earlier. It is arguably the most aerodynamically designed hunting weapon of all, capable of being directed at a specific target but should it miss its design remains in close partnership with the air and returns it to the thrower.

1877 - 1853: Cayley, George - English

A wealthy landowner, made a study of the attributes required to provide lift and sustain flight. His gliders incorporated curved and tilted wings (dihedral), offering a predetermined angle in flight which he regarded as fundamental, and to which he added a rudder. In 1853, he flew his first completely successful model glider and built the world's first glider capable of carrying a passenger. After this event, he published his research notes, which laid the foundations of modern aerodynamics. He is, even today, regarded by many historians as the actual inventor of the aeroplane. (As his studies preceded the I.C.Engine it may be more accurate to claim that he discovered the principles of flight.)

1848 - 1896: Lilienthal, Otto - German

The most innovative of his era building bigger gliders and succeeding in flying to heights which allowed him to study thermal air currents the results of which he released in 1894. He was able, from practical air-borne experience, to describe the influences exhibited when the angles, contours and curvatures met a variety of air pressures. He was the first to achieve non-powered flight and inspired others to master the technique themselves. It was reported that in 1932, one of his students, named Groenhoff, flew a distance of over 160 miles and Dittmar, another German enthusiast, is reputed to have set an altitude record of more than 12,500 feet in 1934. Lilienthal's success with gliders was an added inspiration for the Wright Brothers to whom, it is said, that he was able to proffer much advice.

The Wright Brothers, Orville and Wilbur, spent most of their lives learning as much as possible about gliding and building gliders. Additionally they were both engineers which made them able to understand the Internal Combustion Engine and how it could be adapted to achieve powered flight. Glider knowledge, valuable though it was, had limitations. The air-frame of the machine they were building would have to carry the burden of an internal combustion engine. A method had first to be devised to securely install an engine which would also have to resist the pull of a propeller while converting the power of the IC engine into the motive force to lift the plane off the ground and sustain flight. They worked from 1896 until December 17th 1903 before they successfully achieved a flight of 852 feet and within two years they had increased the distance flown to 20 miles and without any doubt they had changed the world.

1877 - 1958: Alliot V. Roe - England

In 1908, in an aircraft of his own construction, with the exception of the engine, reputable witnesses, testified that Roe did in fact become the first ever to achieve flight in Britain and, in July 1909, he circuited London in a tri-plane, with an engine which was only a 9 horse-power J.A.P. In 1910, he constructed, and flew, his own biplane. He founded a company and one of his planes served as a training machine and another as a bomber in WW1. His company was later reinforced by the addition of Roy Chadwick, one of the most gifted English aircraft designers, who later designed the finest long-range bomber of the Second World War the "Lancaster" (The purist would point out that it began life as the "Manchester" bomber but, with only two engines, it was underpowered. Two more were fitted which then produced the transformation).

During the 1920s, there were international air races for the Schneider Trophy. The main competition was between America and Britain and, initially, America dominated with their military aircraft reaching speeds of 145mph. in 1923 and 232mph in 1925. Then Britain won it three times in succession, winning it outright in 1931, with the "Supermarine". In the last race, a speed of 407.5 mph. was recorded.

1895 - 1937: Mitchell, Reginald J - English

Mitchell designed the brilliant "Supermarine" seaplane and later the legendary "Spitfire", so outstanding in WWII, which was enhanced by the equally brilliant Rolls-Royce Merlin engine. He was a world leader in aircraft design and his early death robbed Britain of an outstanding figure.

1907 - 1996: Whittle, Frank - English

Probably the greatest aero-engineer of the 20th. century, Whittle took Britain and the world into the jet-age. In 1928, he submitted to the War Office a revolutionary paper describing Rocket propulsion and gas turbines, which was dismissed as impracticable by their scientific advisors.

His application for a patent in 1930 was finally granted in 1932 but, he met with total refusal by the War Office to help with development costs and they refused finance to renew the licence when his patent expired. He joined two former RAF pilots to form a company, Power Jets, to keep his invention alive. Considerable publicity surrounded this invention for a number of years but the patent was never subjected to the official secrets act and it was unsurprising that, in 1936, it was learned that the Germans were also embarking on an identical project.

The outbreak of war panicked the Air Ministry who, in 1939, belatedly took a limited, but still parsimonious, interest and it was not until May 1941 that they agreed to supply a plane in which to test the engine.

When the test finally took place, the military chiefs were so impressed by its power that that they immediately made it a top priority and they assumed full control over its development. From then onwards, it appears, that Whittle ceased to have much part in the proceedings and his erstwhile detractors began selecting who would get the contract to manufacture the engines. In the meantime, he was awarded a senior RAF rank. Full details of the jet engine were released to the American government, who appointed General Electric to manufacture them. In 1942 when they encountered difficulties, The War Office sent Whittle to the USA to overcome the problems which is probably why the American version was airborne months before the British Meteor. After the war, Whittle's engine powered the 'Comet', the first passenger jet, the American Boeing 707 and, finally, the fastest passenger jet ever built- the Concorde. Subsequently Whittle went to live and work in the USA.

1910 - 1999: Cockerell, Christopher - English

Invented in 1955, and perfected 1959, the hovercraft was a vehicle designed to travel on a cushion of air, enabling it to traverse over deep or shallow water, sand or marshland and it is now in use throughout the world. The air pressure providing the lift was contained by a skirt of neoprene, stitched by stainless steel wire to the main body of the vehicle. Because the hovercraft was not in direct contact with the Earth, it was regarded in the same category as a plane, so the person at the controls was a pilot.

1919 - 2010: Goodhart, Nicholas - English

When faster and heavier jets were brought into service for the Fleet Air Arm, the incidence of accidents when landing on aircraft carriers increased, Pilots were finding it difficult to maintain an accurate flight-path, especially those with slower throttle response. To overcome this, Goodhart devised a method of placing a concave mirror on the port side of the deck, flanked by green lights. A powerful beam was then directed toward the mirror, creating a ball of orange light which the incoming pilot used as the central point between the green lights, to ensure a secure and safe glide path. It proved so effective in reducing accidents, that the Americans installed the system immediately in their own fleet. They also awarded Goodhart the U.S Legion of Merit in recognition of his achievement.

Some heroes of aviation and a short list of firsts:

1903 Wright Bros. achieved first ever powered flight

1908 Alliot Roe achieved flight (75ft.) with his first plane

1909 Louis Bleriot crossed the Channel in his monoplane (powered by 25hp Anzani engine)

1910 Charles Rolls – first to fly non-stop across the Channel and back

1910 A.V Roe began manufacturing planes-AV504trainer/bomber in WW1

1913 Nesterov & Pegoud inaugurated aerobatics by flying the loop the loop

1919 John Alcock & Arthur Brown(pilot /navigator flew the Atlantic

1926 Alan J. Cobham flew to Cape Town and back and repeated feat to Australia

1927 Charles Lindberg flew solo from New York to Paris. Time 35.5 hours

1930 Amy Johnson-first woman to fly solo from England to Australia, time 20 days

1931 Francis Chichester first to fly Tasman sea East/West to Australia

1935 Amelia Earhart flew from New York to Ireland solo

7. A CHRONOLOGICAL CHART OF SCIENTIFIC PROGRESS

The inclusion of Gutenberg might be considered to have been more appropriate in the inventions section but considering the contribution that came from the dispensation of scientific information, by the printed word throughout the centuries, it seemed apposite.

1214 - 1294: Bacon, Roger - English

A Franciscan friar credited with being the founder of experimental science, optics, chemistry. His profound knowledge of astronomy enabled Bacon to point out inaccuracies in the calendar. It has been claimed that his calculations were later found, by astronomers using more advanced equipment, to be almost exact He invented gunpowder but as an explosive mixture, not a propellant. Ecclesiastical authorities disagreed with him dabbling in alchemy as a result of which he suffered years of imprisonment.

c.1410 - 1468: Gutenberg, Johannes - German

In 1448, Gutenberg is recorded as starting a company in Mainz (or Mayence) and engaged in developing his invention of the printing press.

In common with most burgeoning companies he sought financial investment presumably to buy the raw materials to cope with the growing demand for his printed books together with the additional employees he would require.

There was also another motive. He had now decided to replicate, by mechanical means, ecclesiastical works of the highest artistic quality which for centuries had been done painstakingly by the skilled hands of monks. He had already printed the Mazarin, (or 42 line) bible but now his ambition was to create an ornate Psalter. The costs in the development stages of all prototypes are difficult to forecast and this might have been another reason that he required additional finance.

His most skilled employee, Peter Schoffer suggested that his own father-in-law Johann Fust (or Faust) might be a suitable investor. Gutenberg and Fust agreed that in return for a loan Fust would receive a partnership in the company. The Psalter was near completion when Fust suddenly demanded immediate repayment of the loan and a legal dispute ensued. The court ruled that Gutenberg forfeit his company in lieu of repayment and Fust, the new owner, then put on sale the beautiful ornate Gutenberg Psalter which, in 1457, realised a greater sum, by many times, than the total worth of the company.

1473 - 1543: Copernicus, Nikolaus - Polish

Generally regarded as the founder of modern astronomy, the fame of Copernicus rests principally on his study and knowledge of the universe His statement that the Sun was the centre of a system around which the Earth and other planets revolved, caused a certain amount of consternation as it was seen as contradicting current belief.

1500 - 1617: Napier, John - Scottish

Napier was a mathematician, who developed the concept of logarithms as a calculating device. It was applauded by scientists and mathematicians everywhere, as was his other device for measuring by rods. (Napier's bones). Napier also devoted much time to religious crusading.

1584 - 1642: Galileo, Galilei - Italian

Galileo invented the first astronomical telescope, offending religious leaders with his scientific assertions, who believed he was defying religious dogma. Under torture, he was forced to deny his belief that Earth orbited the Sun. It was surprising, when in another country, a century earlier, Copernicus had made the same statement encountering only grudging acceptance.

1571 - 1630: Kepler, Johann - German

Kepler was a great mathematician and astronomer and one of his major works was published as Keplers Laws, describing his theory of the planets. This work included:-
1. Every planet describes an ellipse, the Sun occupying one focus.
2. The radius vector (line joining the centre of the Sun to the centre of the planet) of each planet sweeps over equal areas in equal times.
3. The squares of the periodic times (the periods of complete revolution round the Sun) of two planets are proportional to the cubes of their mean distances from the Sun. It has been suggested that the observance of these laws assisted Newton in his law of gravitation.

1602 - 1686: Guericke, Otto von - German

In the middle of the 17th century, Guericke demonstrated how static electricity could be produced by the application of friction, in effect the first electrical generator. He also proved that light travelled through a vacuum, whereas sound did not.

1627 - 1691: Boyle, Robert - Anglo / Irish

Boyle was a leading figure in the study of gases, publicised as Boyle's Law. He was an early member of the Royal Society and, from his scientific studies, he delivered a theory of the elements.

1631 - 1723: Wren, Christopher - English

Wren's early reputation was built on his profound knowledge of astronomy and he was the Savilian professor at Oxford. He was appointed by Charles II to restore the old St. Paul's but the Great Fire in 1666 made it necessary to re-build it completely. He went on to design the library of Trinity College, Cambridge, the modern part of Hampton Court Palace, hospitals and he also built fifty two London churches.

1635 - 1703: Hooke, Robert - English

A man of eclectic thought and ingenuity, Hooke subjected all manner of materials to the effects of stresses and strains, enabling him, in 1678, to define the laws of elasticity and the observance of these laws is present in all branches of applied mechanics. His contribution to astronomy was considerable and, to further these studies, he built a powerful telescope. In 1664, he became the Professor of Mechanics to the Royal Society and, as evidence of his genius, he invented a device able to transmit a continual motive force between two shafts, even though they may not be entirely parallel. Today, that invention is used in modern day motor car manufacture and known as the 'universal joint'. Hooke was also responsible for the rebuilding of London after the Great Fire.

1642 - 1727: Newton, Isaac - English

Widely regarded as possibly the greatest genius of all time, Newton specified the laws of motion and gravitation. His work in pure mathematics, optics and the nature of the spectra, together with the discovery of the principle of calculus, is evidence of his originality and vision. The result of his knowledge of optics allowed him to invent the reflecting telescope. He also explained the forces governing planetary behaviour and his great work, Philosophe Naturalis Principia, was published in 1686 and 1687.

1656 - 1742: Halley, Edmund - English

Halley correlated the comets with the solar system and predicted that Halley's comet would return to the Earth's atmosphere at given frequencies. In 1682 his prediction of this event was correctly given as 1789. He spent two years (1676-1678) on St. Helena from where he catalogued the stars and constellations of the southern hemisphere.

1682 - 1744: Hadley, John - English

Hadley invented the quadrant originally intended for calculating the altitude of the stars but proved to have practical uses in navigation, surveying and gunnery, as well as furthering the science of astronomy. He subsequently invented the sextant.

His improvements to the reflecting telescope invented earlier by Newton might have triggered some sensitivity, because Newton then claimed part of the credit for the sextant because of information he had supplied.

1685 - 1768: Hadley, George - English

Central to George Hadley's studies were the effects of weather patterns on meteorology, in particular the influence of the trade winds on atmospheric conditions. He was a member of the Royal Society and much of his work was explained in his papers, "The Hadley Cell".

1686 - 1736: Fahrenheit, Daniel - Polish

Fahrenheit's claim to fame rests mostly on his invention of the first thermometer. In his first one he used alcohol to register the reading on the temperature scale but in 1714 he changed to mercury. He also excelled in the making of precision instruments for the study of meteorology.

1693 - 1776: Harrison, John - English

In the early part of the 18th century, all maritime nations were concerned at the high number of losses of ships, mariners and their valuable cargoes. Under adverse conditions, navigators could only calculate the latitude - not the exact longitude. An Act of Parliament passed in 1714 offered rewards to anyone who could overcome the problem. For an accuracy of within 60–40–or 30 nautical miles, the prizes were respectively £10,000--£15,000 & £20,000. These large amounts indicated the scale of the problem. The greatest astronomers could not provide a solution and Harrison, by his invention of the chronometer, won the prize and revolutionised marine navigation. Now mariners were able to calculate their true position at sea, whatever the conditions or visibility. In the course of his work, he mastered the principles of the differing expandability of metals and invented the fusee which allows a watch to be wound without interrupting its movement. The mystery is how? -He was not a watchmaker, scientist or a metallurgist but a carpenter, yet he incorporated all of these skills in his invention.

1700 - 1745: Kleist, Ewald - German
c.1701 - 1775: Mussenbroek, Pieter - Netherlands

In the Netherlands in 1744, a simple experiment took place by Mussenbroek of Leiden University. A jar was partly filled with water, the neck was then blocked and a wire was inserted into the jar until it reached the water. It was subjected to a small amount of friction-based static electricity and showed signs that it was capable of storing a small charge of electricity. It was then discovered that Kleist had conducted the same experiment the year before, with a similar conclusion. Mussenbroek never studied it further and Kleist died the following year, so the idea stagnated – until William Watson, an English scientist, believed it to be worthy of further development and began his own experiments.

By 1747, he had developed its capacity to store static electricity substantially and was able to demonstrate the transmission of electricity over considerable distances. What Watson called the Leyden Jar eventually became known as a capacitor (or condenser) and, centuries later, is now an essential component of many electrical devices, including radio and television receivers.

1701 - 1744: Celsius, Anders - Swedish

In 1742, Celsius designed a thermometer but he took for the base calibration 0 degree as the freezing point of water and continued in equal parts up to 100 degrees which he designated as the boiling point of water.

1706 - 1790: Franklin, Benjamin - American

Apart from a busy life as a publisher, Franklin wrote extensively about electricity and its effects. One of his experiments was to test the action of lightning and, during a storm, he used what he called a" lightning rod" and, in effect, proved the principle of "earthing", an electrical charge. His experiment had demonstrated that by mounting a metal rod on top of a building, it channelled the charge safely down to the Earth. During one of his dangerous experiments he was lucky to escape with his life.

1728 - 1799: Black, Joseph - Scottish

He performed experiments on various gases and the effects of subjecting them to differing temperatures. His work influenced James Watt, to whom he also acted as mentor.

1731 - 1810: Cavendish, Henry - English

Cavendish published some papers including, in 1801, the measurement of the mass of the Earth but, being a recluse, he conducted most of his work in isolation. As a result, many of his findings were not revealed until after his death, when James Clerk Maxwell undertook the study of notes and manuscripts which he had left. Among them he discovered that Cavendish had anticipated many of the discoveries credited to others, amongst them, Faraday, Coulomb, Ohm and Black. The prestigious Cavendish laboratory at Cambridge was inaugurated in his honour.

1733 - 1804: Priestley, Joseph - English

His book, "The History and Present State of Electricity", which was published in 1767, dealt with additional aspects of the Leyden Jar. He discovered oxygen and a method of obtaining gases, collecting them by using a pneumatic trough over mercury. The eight gases he was able to garner in this way were hydrogen, hydrochloric acid, nitrous oxide, ammonia, carbonic oxide, sulphur dioxide, nitric acid and silicon tetrafluouride. Vilified for supporting the French Revolution, he eventually emigrated to America.

1736 - 1806: Coulomb, Charles A - French

Coulomb investigated many aspects of the effects of electrical charges, including that of polarity, resulting in a law named in his honour. He also built special apparatus to help prove the laws of electrical forces in practical experiments.

1737 - 1798: Galvani, Luigi - Italian

A university lecturer and teacher, Galvani made a particular study of animal physiology and, during the dissection of a frog, he noticed that when contact was made between the leg of the frog and a metal bar, it caused a twitching reaction in the animal's muscles. He claimed it was due to the innate presence of another force within the body which was stored in the brain (animal electricity) and operated between nerve and muscle, excited by bridging them with metal

1745 - 1827: Volta, Alexander - Italian

Although Volta disagreed with Galvani's theory, it had captured his interest and led to a series of experiments. He rejected the idea of using the leg of a frog and found that by using copper and zinc plates separated with moist wadding or paper it produced the same chemical reaction and when the same items were immersed in a jar of diluted sulphuric acid it was transformed into electrical energy. His later experiments included zinc, silver and other metals. From his conclusions he was able, in 1800, to demonstrate to the scientific world the first ever electric battery (or Voltaic pile) designed to deliver continuous current.

1753 - 1815: Nicholson, William - English

Nicholson was a combination of experimental chemist and mechanical inventor. He invented a hydrometer for measuring the density of liquids. In common with most scientists he took great interest in the Volta battery which he incorporated into some of his own experiments. He immersed the terminals from a battery into water and discovered the process of electrolysis of water, a procedure now common to electro-chemical research.

1755 - 1833: Elhuyar, Fausto - Spanish

The mineral initially known variously as scheelite (discovered by W.Scheele), or wolfram. The name was changed to Tungsten when it came into the hands of the eminent scientific chemists and mineralogists, Fausto and his brother Juan Jose. They were the first to isolate the original mineral and discovered properties in it which they used as an alloy to enhance the hardness and quality of steel. They dedicated most of their lives to teaching chemistry and metallurgy.

1766 - 1828: Wollaston, William H - English

Wollaston devised some original metallurgical techniques for purifying and processing, in particular, molybdenum and platinum. The latter was generally found to be difficult to produce in quantity but became a particular favourite of his because of a special procedure he had evolved. It was a method which it was said he never divulged because the income it regularly generated sustaining him throughout his later life. In the first decade of the 19th. century, he also discovered palladium and rhodium.

1766 - 1844: Dalton, John - English

Dalton developed the atomic theory of matter and is considered one of the founders of physical science. In 1808, he presented his "New System of Physical Science", which made him a universally respected scientific figure. From his studies he developed the atomic theory of matter.

1773 - 1829: Young, Thomas - English

Studying optics, Young discovered the condition astigmatism. His research on wave-light theory offered a calculated measurement of the seven colours, as designated by Newton. His discovery of the interference of light and wave phenomena was rejected by British scientists, as being contradictory to Newton but a French scientist supported and conclusively proved his theories : See Fresnel 1788-1827.

1775 - 1836: Ampere, Andre M - French

Following wide-ranging experimentation, Ampere founded the science of electrodynamics. He formulated a mathematical way to measure between electrical currents and built an apparatus for calculating it- the ammeter. Ampere's name is used to describe a unit of electric current.

1776 - 1862: Barlow, Peter - English

Barlow invented the achromatic lens, which eliminated colour distortion. He also studied the adverse effect on ship's compasses which had been experienced since the introduction of iron hulls. He corrected the problem by positioning, next to the compass, an iron object shaped so that it counteracted the deviation. He received a commendation from the Royal Society for his intervention.

1777 - 1851: Oersted, Hans C - Danish

In 1820, Oersted by the discovery that a magnetic needle could be deflected by an electric current was reputedly the first to prove the co- relationship between electricity and magnetism.

1778 - 1829: Davy, Humphrey - English

Numerous explosions were caused by a build-up fire-damp in mines and, to overcome the problem, Davy invented a safety-lamp which saved many lives. In his researches, he also discovered sodium, potassium, calcium, chloride and "laughing gas".

1783 - 1850: Sturgeon, William - English

Engaged in electrical experimentation, Sturgeon succeeded in producing the first practical electro-magnet in 1825 and this had a bearing on the invention of the telegraph. He also built an electric motor and, in the course of construction he invented the Commutator, a device fitted to most modern electric motors, which changes the current from alternating to continuous (or direct).

1786 - 1853: Arago, Francois - French

Arago's experiments showed that magnetism could be produced by the rotation of a non-magnetic conductor. By passing an electrical charge through a spirally-formed length of copper wire, he found that, for the duration of the charge, it was able to pick up iron filings similar to any normal magnet. In these experiments, he was building on the work of Oersted.

1787 - 1854: Ohm, Georg S - German

Ohm was the originator of the law of the theory of voltaic current. Known as "Ohm's Law", an important law of electricity it is used to describe a unit of electrical resistance. In order to accurately measure resistance, he created an instrument, unsurprisingly named an Ohmmeter.

1788 - 1827: Fresnel, Augustin J - French

Further to theories expressed by Young, Fresnel enlarged them by establishing his undulatory theory of light. He put this into practice by surrounding a central lens with prismatic glass rings, restricting any escape of light and increasing the intensity by a number of revolving lens, thus producing an apparatus for lighthouses which emitted a ray far exceeding anything in use at the time and is still in use today.

1790 - 1845: Daniell, John F - English

To measure humidity, Daniell invented the hygrometer and published substantial information on meteorology and the subject of artificial climate. As professor of chemistry at King's College, he discovered a method of making a voltaic cell that was capable of maintaining a continuous and powerful electrical current. This became known as Daniell's Cell.

1791 - 1867: Faraday, Michael - English

Often referred to as the Father of Electricity, in 1881, Faraday built the first generator and later his discovery and publication of the laws of electro-induction and magnetism brought him world renown. His theory of the laws governing electro-chemistry and induction deduced from his researches into electrolysis also consolidated his position.

1791 - 1872: Morse, Samuel F. B - American

Educated at Yale College, Morse came to England in 1811 to study painting and succeeded to a level which earned him a gold medal from the Royal Academy. He became interested in telegraphy possibly from hearing about the experiments being made by Wheatstone & Cooke while in England. After returning to America he devised a system which used the electric telegraph to send electric impulses which were converted into dots and dashes in sequences that equated to letters. He applied to the US government to accept his system of communication and in 1843 the US Congress granted him leave to install it for a trial period. It proved so successful that it was accepted immediately and by 1857 it was also accepted by European countries and known as the Morse Code.

1801 - 1868: Plucker, Julius - German

In 1858, Plucker inserted two electrodes into a Geissler tube, expelled the air and then subjected the electrodes to an electrical current. The tube became luminescent, due to rays emanating from the cathode. The obvious connection here is with television, which exists because of the of the Cathode tube studies. Together with the Vacuum Tube they provided great impetus to the development of electronics especially computers. (see - Geissler 1815-79)

1802 - 1875: Wheatstone, Charles - English

Professor of philosophy in Kings College London, in 1834, Wheatstone demonstrated the velocity of electricity and its application to telegraphy. In 1837, in co-operation with Cooke, he patented the first electric telegraph. Other inventions included a device for measuring actual resistance, capacitance, frequency and inductance, as well as the concertina, the electric clock and the stereoscope.

1806 - 1879: Cooke, William F. English

Cooke and Wheatstone formed a successful partnership, culminating in the invention of the first electric telegraph, which was also the first telegraphic system to be installed on railways.

1809 - 1882: Darwin, Charles - English

In 1831, Darwin joined a surveying voyage on H.M.S. Beagle, which was commanded by Captain Fitzroy. During a circumnavigation of the globe, he observed that animals of the same species inhabiting islands hundreds of miles apart had evolved physical differences. His book, "The Origin of the Species by Means of Natural Selection" explaining the evolutionary process was considered by many to oppose biblical teaching, publication was delayed until 1859.

1811 - 1899: Bunsen, Robert - German

By a meticulous study of the elements, Bunsen was able to detect that each of them emitted light with an individual wavelength, important to science generally and particularly to astronomers studying the Sun and other astronomical bodies. The Bunsen burner was also of his creation.

1815 - 1879: Geissler, Heinrich - German

By trade a glassblower and in, 1854, while assisting Professor Julius Plucker a German physicist, made improvements to the "vacuum tube", sometimes known as the "Geissler Tube". It heralded the beginning of investigations into cathode rays.

1818 - 1889: Joule, James P - English

Joule studied with Dalton at Manchester and introduced "Joules Law",the first law of thermo-dynamics describing the connection of heat between mechanical and electrical properties. He worked in partnership with the great Irish scientist, Lord Kelvin, studying gases to develop refrigeration and other forms of energy.

1822 - 1895: Pasteur, Louis - French

A biologist chemist and bacteriologist, Pasteur's research into the part played by microbes in decomposition and fermentation enabled him to treat diseases in cattle and discover an antidote for hydrophobia by means of inoculation. He also developed a process known as pasteurisation (not to be confused with sterilisation) for the treatment of cow's milk, which eliminated the injurious bacteria associated with tuberculosis.

1824 - 1907: Thomson, William (aka. Lord Kelvin) - Irish

Physicist, mathematician and inventor, Thomson's knowledge of tidal shifts enabled him to provide valuable advice in 1866, when the Atlantic cable was laid. Thomson's skills, both theoretical and practical, were wide-ranging. He was reputedly the foremost scientist of his time, covering thermodynamics, electricity, heat, elasticity and hydrodynamics. His contribution to the development of radio broadcasting was also of great value because of his knowledge of telecommunications.

1828 - 1914: Swan, Joseph W - English

Swan was the inventor of the incandescent electric lamp, then a miner's electric safety lamp, photo-mechanical printing and electro-metallurgy. He also invented the dry collodium plate for photography. He later joined forces with Thomas Edison, the prolific American inventor.

1831 - 1879: Clerk-Maxwell, James - Scottish

James Clerk-Maxwell revolutionised electrical theory. His electro-magnetic theory of light inspired the experiments which led to the discovery of electric waves which opened up another avenue of scientific exploration.

1832 - 1919: Crookes, William - English

In 1879, Crookes researched the area surrounding the cathode, now known as the "Crookes Dark Space". In 1879, he concluded that the particles emanating from the cathode were negative but could be diverted by a magnetic field from their normal straight course and that the luminescence was part of the electrical current. He showed also that the type of gas used or type of metal of the electrode employed was immaterial, as the luminescence remained. He also discovered the element thallium.

1805 - 1865: Lenz, Heinrich - Russian

Issued his findings on polarity known as Lenz' Law in 1834

1836 - 1920: Lockyer, Joseph N - English

In a spectroscopic study of sunspots in 1866, Lockyer found disturbances or upheavals in a layer around the Sun, which he called the chromosphere. In 1868, when studying the Sun's atmosphere, he discovered a hitherto unknown element which he named helium (coincident with Pierre Janssen, a French astronomer). It was not until 1895 did scientists discover its existence on Earth. (See: Ramsay)

1837 - 1898: Newlands, John A. R - English

An analytical chemist who catalogued all the elements, known at that time, by their atomic weight, Newlands grouped them into seven divisions, each containing individual elements. He later described them in his " Law of Octaves" .

1834 - 1907: Mendeleev, Dmitry - Russian

Similar to, but more extensive than, work done by Bunsen and Newlands. His work in 1871 is of particular importance for advocating the accurate classification of the elements. He anticipated that, over time, there would inevitably be more elements discovered which would need to be added and made the necessary allowances for this. His work became known as "The Periodic Law."

1842 - 1919: Rayleigh, John W. S - English
1852 - 1916: Ramsay, William - Scottish

Eminent scientists who enjoyed a successful partnership, Rayleigh was the professor of Experimental studies at Cambridge and William Ramsay, a chemist and a teacher. They discovered and isolated argon and neon in 1894 and, four years later, added krypton and xenon. Rayleigh also published a much acclaimed book on the subject of acoustics in 1877. In 1904, each received a Nobel Prize .

1850 - 1925: Heaviside, Oliver - English

Heaviside predicted that there was an upper layer of the atmosphere which deflected radio waves from continuing into space and caused them to follow the curvature of the Earth. It was many years before his theory was verified. (see Appleton)

1851 - 1940: Lodge, Oliver J - English

One of the first to demonstrate the possibility of communication by wireless telegraphy, Lodge is credited with perfecting the clarity of this method of communication. He also invented a radio wave detector.

1852 - 1908: Becquerel, Antoine H - French

Conducting studies into uranium, Becquerel discovered radioactivity. He later joined forces with Pierre and Marie Curie to whose work he made a significant contribution. In 1903 they were collectively awarded the Nobel prize for physics.

1856 - 1940: Thomson, John J - English

Thomson confirmed the cathode rays as a stream of negatively-charged particles and developed the theory that electricity and magnetism were inter-related which gave added impetus to the study of the cathode.

In 1897, his research showed the atom was more complex than previously thought and realigned the direction of research; atomic structure, he discovered, comprised smaller particles of the same type, which he called electrons.

1857 - 1894: Hertz, Heinrich - German

Hertz pursued the theories of Clerk-Maxwell and became the first to broadcast and receive radio waves.

He also proved that light and heat are magnetic radiations.(see Radar section).

1858 - 1947: Planck, Max - German

Max Planck is principally recognised as the originator of the Quantum theory, which added to and influenced much of 20th. century understanding of atomic structure

1859 - 1906: Curie, Pierre - French
1867 - 1934: Curie, Marie - Polish

Remarkable scientists who together achieved outstanding results. In 1903 and in co-operation with Bequerel they were Nobel prize winners for physics. Marie discovered the first radio-active substance, which she named polonium, in honour of her birthplace. They then concentrated on radium, which they discovered in pitch-blend. In 1911 Marie personally won the Nobel prize for chemistry.

1871 - 1937: Rutherford, Ernest - New Zealand

In 1919, Rutherford made great scientific progress on the subject of nuclear reaction, discovering a reaction between a nitrogen nucleus and an alpha particle, subsequently concluding that progress could be advanced only by accelerating particles to extremely high energies. His forecast was vindicated when, by this process, it was successfully achieved by J. D. Cockcroft and E.T. Walton in 1932.

1874 - 1937: Marconi, Guglielmo - Italian

In 1901, Marconi succeeded in transmitting a radio signal from Poldhu in Cornwall to St. John,

Newfoundland. His discoveries and inventions had an immense impact on radio communication, broadcasting and navigation.

1879 - 1968: Hahn, Otto - German

Hahn studied with nuclear physicists, Ramsay and Rutherford, and discovered nuclear fission. In 1944, he received a Nobel prize for physics.

1879 - 1955: Einstein, Albert - German

One of the world's great scientists. His " Theory of relativity" had a profound effect on scientific thinking.

1881 - 1958: Thomson, George - English
1892 - 1975: Davisson, Clinton - USA

In 1937, Thomson and Davison were jointly awarded a Nobel prize for physics. Their studies of the atomic structure of liquids revealed that electrons can be diffracted in a similar manner to light waves and particles.

1885 - 1962: Bohr, Neils - Danish

In 1922, Bohr received a Nobel prize for physics, in recognition of his work in defining atomic and molecular structure.

1887 - 1894: Hertz, Heinrich - German
1882 - 1964: Franck, James - German

Based on the theory of Bohr, Hertz and Franck's work in physics, electrical experiments and atomic research, earned them Nobel prize awards (see Radar and Radio sections). Hertz also built an obstacle detector, based on the electro-magnetic theories of Faraday and Maxwell and the potentialities were keenly observed and developed by others.

1892 - 1965: Appleton, Edward - English

Appleton was recognised for his research of electro-magnetic waves and, In 1947 he discovered the exact location of the layer in the ionosphere which dependably reflected radio waves back towards Earth.

1897 - 1967: Cockcroft, John D - English
1903 - 1995: Walton, Thomas S - Irish (Northern)

Cockcroft and Walton built a particle accelerator to study atomic particles from which they developed the Cockcroft generator. In 1932, they announced that they had split the atom and became joint winners of the Nobel Prize in 1951 for their work.

1913 - 2012: Lovell, Bernard - English

Lovell was an important member of a scientific research team set up by the government during WW2 to develop RADAR. After the war, he returned to lecturing at Manchester University, simultaneously constructing at Jodrell Bank, what became, by 1957, the world's biggest radio telescope. Russia and America were racing to be first to launch a satellite into space and the jubilant Russians won with Sputnik in October 1957 but then realised that they lacked the technology to locate its position during orbit. The only man on the planet who could tell them its location was Bernard Lovell, by using the equipment he had scrimped and scrounged to build. Later, when the American satellite was launched they were faced with the identical problem and he was requested by their military command to provide them with the same service.

8. ENGINEERING INVENTIONS AND METALLURGY

c.287 - 212 BC: Archimedes - Greek

Probably the most quoted of the ancients to leave a legacy of practical work.

The Archimedes Principle is an example of his analytical mind and as taught to every schoolchild. (that the weight of an object immersed in a fluid is equal to the weight of the displaced fluid) It was by observing the 'Principle' he was able to prove that a crown made for King Hiero, although supposed to consist of pure gold, fraudulently contained a lighter alloy.

1599 - 1684: Dudley, Dud - English

The production of iron for military purposes was a prerequisite in the 16th and 17th centuries. Smelting methods had not changed for centuries and the use of charcoal was denuding the forests alarmingly. Dudley converted to the use of coke which created greater heat. He was granted a patent for the process in 1621. (Incidentally the name Dud did not reflect the view of a disillusioned parent but was also the name of am 8th century prince of West Mercia.)

1623 - 1662: Blaise, Pascal - French

A child prodigy in geometry, languages, physics and philosophy, Blaise invented the first calculating machine in 1647 . Surprisingly, it took almost 250 years before another inventor improved on his model (see W.S.Burroughs).

1675 - 1717: Darby, Abraham - English

Darby also used coke his larger furnaces and produced iron of such fine quality that he was able to make cylinders and pressure boilers for Newcomen and Trevithick. It was a family business, his grandson produced the world's first iron bridge, now a national monument and a design system later used by Telford.

1706 - 1788: Boulsover, Thomas - English

Boulsover invented the method of fusing silver onto copper, creating what became known as "Sheffield Plate" in 1742. This discovery enabled the burgeoning middle class to own tableware which, to all appearance, was the equal of solid silver. Boulsover also invented the high carbon steel saw blade.

1716 - 1772: Brindley, James - English

Brindley was an innovative builder of canals aqueducts and bridges. In 1765-70, in order to complete the Bridgewater canal, he incorporated an aqueduct which spanned the River Irwell at Barton, Lancashire enabling coal to be transported into the centre of Manchester. Other works of his were, the Grand Trunk Canal to link the Trent and Mersey and later uniting that with the Severn.

1718 - 1794: Roebuck, John - English

Chemist, inventor and entrepreneur, Roebuck founded the Carron ironworks near Falkirk in Scotland, a company near Edinburgh for bleaching cloth and he also invented the mass production of sulphuric acid. He needed a more powerful steam pump to clear water from deep mineshafts, so financed Watt to study the Newcomen model.

1724 - 1792: Smeaton, John - English

Smeaton's revolutionary design for lighthouse construction used interlocking dovetailed concrete blocks, imparting immense strength and solidity. The binding agent was hydrated lime and powdered grit, which set under water. One of his other great works was the Forth & Clyde canal, 35 miles in length, linking the North sea and Atlantic ocean (see Aspdin).

1728 - 1809: Boulton, Mathew - English

With factories in Birmingham, Boulton was a major industrialist and inventor and, after the bankruptcy of John Roebuck, he agreed to continue the financial support for James Watt, enabling the improved steam engine to be completed. In addition, Boulton provided skilled engineers and all the facilities to bring the modified steam pump to completion. In 1775, they formed a company, called Boulton & Watt (see entry for Roebuck).

1740 - 1800: Cort, Henry - English

In 1783, Cort patented machinery for producing iron by "puddling" and then of reducing the thickness of iron bars by rolling them, instead of laboriously hammering them. Both of his inventions increased production and reduced costs. The government of the time acknowledged his contribution by granting him a pension in 1794.

1749 - 1814: Bramah, Joseph - English

Excelled with inventions such as banknote printing machines, thief-proof locks and a hydrostatic press, Bramah consolidated his reputation in the development of the machine tool-making industry. Some of his employees having benefitted from his tuition became great inventors in their own right, including Maudsley, Whitworth, Wilkinson and Naismith.

1757 - 1816: Greathead, Henry - English

In 1789, Greathead patented the first lifeboat capable of negotiating rough sea conditions. Its success was such that an appreciative government was prompted to award him £1,200.in recognition.

1757 - 1834: Telford, Thomas - Scottish

A builder of canals and roads including the Ellesmere canals and a network of canals in Sweden. Many roads in Scotland were constructed under his supervision. In 1826 he also supervised the construction of the first Menai bridge in North Wales.

1755 - 1883: Elhuyar, Fausto - Spanish

Both scientists, Fausto and brother Juan took an interest in the element tungsten. After isolating it they found that, when allied to steel, it proved ideal for making high-speed cutting tools and later, filaments for light bulbs (NB. tungsten and wolfram are synonymous – Wolfram was discovered by Scheele in 1781 but he took no further interest in its possibilities). {Perhaps it is speculation but they might have been directly responsible for the famous Toledo steel.}

1760 - 1820: Lytton, John - English

In 1794, Lytton invented an early design for a propeller which incorporated the "Archimedes Screw" principle but dropped the project and his patent lapsed. This was later revived by Ericsson, with the introduction of iron-clad ships(see Ericcson entry).

1768 - 1865: Donkin, Bryan - English

Donkin invented a prototype of the rotary press and other improvements to machinery for paper-making. Another of his inventions was the composition roller, which came into extensive use within the printing industry.

1771 - 1831: Maudsley, Henry - English

The machine tool industry is indebted to Maudsley. He invented the earliest machine tool, the metal turning lathe and other attachments, making it possible to manufacture other machines to perform different functions, including drilling, boring and automatic screw-cutting. He trained numerous engineers and, during the Industrial Revolution, some became famous inventors in their own right.

1786 - 1865: Hancock, Thomas - English

In the 15th. century, rubber, also referred to as cahuchu, was discovered in South America in plentiful supplies but, as nobody could offer a way of converting it from its natural state, it lacked investors. Until in 1821 when Hancock invented a machine which he called the "Masticator", which shredded the material into an emulsified mass that was then rolled into sheets similar to cloth. By this means, he founded the rubber industry

1766 - 1843: Mackintosh, Charles - Scottish

Mackintosh learned of Hancock's invention and believed the sheets could be made into waterproof coats. In 1822, he applied for a patent under his own name and the two men formed a partnership to make garments which they called "mackintoshes" but the project failed, because the stitching leaked. Success was guaranteed when another inventor devised a cold liquid adhesive, called Parkersine, which was capable of forming an effective watertight seal between layers of the material. (see entry for Alexander Parkes).

1769 - 1849: Brunel, Mark Isambard - French

Working in America before coming to England, Brunel, a gifted engineer, soon found himself a niche in industry. His most important contribution was the invention of the "tunnelling shield", by means of which he achieved the first ever subaqueous tunnel. Opening in 1843, this went under the Thames from Rotherhithe to Wapping.

1790 - 1860: Aspdin, Joseph - English

Aspdin is considered to have produced the first true cement. Patented in 1824, and because of its appearance being similar to Portland stone, he named it Portland cement. He added silicon to the mix, which imparted greater adhesive power, even under wet conditions.

1803 - 1887: Whitworth, Joseph - English

The originator of the process of "fluid pressed steel" for producing cannon and ships' plates, Whitworth also invented the "Whitworth gun", a device which could measure to a thousandth part of an inch and a method of scraping iron surfaces flat, to that limit. His screw-cutting lathes could churn out standardised precision bolts and nuts in immense numbers. At the 1851 Great Exhibition, his site occupied almost a quarter of the available space and, for the first time, engineers could buy pre-made nuts and bolts, each cut with standard Whitworth threads.

c.1780 - 1870: Bakewell, Benjamin - English

In 1808, Bakewell, an English émigré, built a factory in Pittsburgh to manufacture cut glass products of the highest quality and of exquisite design. Bakewell was regarded as the Father of US Glass industry.

c.1800 - 1870: Budding, Edwin - English

In 1830, Budding invented the lawnmower, not only a boon to the landscape gardener or the owners of large estates but any householder with a patch of grass. It was now possible to provide accurate surfaces for a variety of sports such as golf, tennis, cricket, bowls and football.

1879 - 1962: Freyssinet, Eugene - French

By his invention of pre-stressed concrete in 1938, Freyssinet advanced construction techniques enormously. He employed high-quality steel to provide strength and stability. His techniques made an enormous improvement to the construction of bridges and came at a time when bigger and taller buildings were being envisaged.

1803 - 1889: Ericsson, John - Swedish

Ericsson came to England and worked for some years on railway locomotive construction. in 1836, he patented the screw propeller, which was first fitted to a ship built in London. The American Naval Department made tests and some years later they and British Admiralty fitted the screw propeller to their war ships.

1806 - 1859: Brunel, Isambard K - English

Brunel was ingenious and innovative and after receiving practical tutoring by his father in 1825 he was appointed chief engineer on the Rotherhithe tunnel project an

assignment on which he almost died. He designed and built the Clifton Bridge which spans the Avon Gorge which, standing 245 feet above the river bed uniting the counties of Somerset and Gloucestershire. He also built thousands of miles of railways in various countries. He designed and built three iron-clad steam-ships. The first, in 1837, was the Great Western and, fitted with wooden paddles, it was the first steam-ship to offer a regular transatlantic service: The Great Britain built in 1843 a steam-ship with an iron hull, the first large vessel driven by a screw propeller. The third was The Great Eastern, equipped with screw-prop and paddles, famous for enabling the first successful Atlantic cable to be laid.

1808 - 1890: Nasmyth, James - Scottish

Nasmyth was a protégé of Maudsley and, after some years, he left to start up a company of his own in Manchester to make machine tools. When Brunel was building the world's first iron ship and required some extra heavy pressing work done, it was Nasmyth who provided the solution by his brilliant invention, the "Steam Hammer".

1810 - 1900: Armstrong, William G - English

Armstrong practiced law, before expressing his ability as an inventor. The hydraulic crane in 1854, and the first rifled ordnance gun were of his invention. He converted dock gates and capstans from manual control to operate hydraulically. His invention of the hydro-electric machine produced frictional electricity. It was driven by a waterfall which he created by linking two minor tributaries of a nearby river which passed through his estate. The electricity generated from his hydro-electric system was channelled into his house and was sufficient to supply all the electrical power to run all his household needs. His house contained one hundred rooms, a lift to all floors, a hydraulically-powered spit, full central heating, an electric dinner gong, a hot water supply and, in the basement, a hot tub. This, the first house in Europe to be fully-lit by electricity and was completed before 1884. It is now owned by the National Trust.

1811 - 1861: Otis, Elisha - American

Otis invented a hoist for lifting heavy items, which incorporated a safety device which guaranteed the security of the load, even if the rope or chains broke. It eventually developed into the world-renowned "Otis Elevator".

1813 - 1898: Bessemer, Henry - English

Bessemer discovered that, by blasting air into molten iron while it was in a state of fusion, steel could be made directly from pig-iron. Patented in1855, it revolutionised the production costs which plummeted resulting in a vast increase in steel manufacture throughout Europe and America. Its importance can be judged by the reaction of one steel producing town in Alabama USA, which named itself Bessemer in his honour.

1813 - 1890: Parkes, Alexander - English

At his Birmingham factory, Parkes produced'Celluloid' the first ever plastic. While experimenting with rubber, a bi-product of latex, he discovered a solution capable of creating a watertight bond between sheets of the material, which he patented under the name of "Parkersine".

1817 - 1898: Fowler, John - English

Fowler was unsurpassed in the construction of great projects:- He assisted in building the London to Brighton railway and apart from acting as Chief engineer on the Stockton and Hartlepool railway line was also, in 1844, the consultant in the construction of the railway between Sheffield and Manchester.
As engineering advisor to the Viceroy of Egypt 1898-1904 and, working to the designs of the English engineer, William Willcocks, he was responsible for the construction of the Aswan Dam in the upper Nile

valley, to provide hydro-electric power for Egypt. This dam was completed in 1902.

Some of his other achievements were in partnership with Benjamin Baker viz:

The design & construction in London of - first underground railway in the world

Design and building of Victoria Station and Pimlico Bridge

The Forth Bridge in Scotland in 1889 a project costing £1,600,000.

1832 - 1923: Eiffel, Alexander G - French

Eiffel was one of the first engineers to use the compressed air caissons in bridge-building. The Eiffel Tower in Paris, and the locks on the Panama canal, clearly demonstrate his ability.

1840 - 1907: Baker, Benjamin - English

Baker was chief designer and, with John Fowler, the joint builder of the Forth Bridge in Scotland which, though completed in 1889 is, even today, a world-wide object of admiration and wonder. (See also- John Fowler)

1828 - 1910: Leader, Williams Edward - English

Leader-Williams was the designer and builder of the greatest of all British canals and the Manchester Ship Canal, which opened in 1894, permitting ocean-going liners to travel the 36 miles inland to the centre of Manchester to deliver raw cotton direct from India, Egypt and America. The Bridgewater Canal, already built across the River Irwell at Barton, was an impediment to the passage of large ships so, to overcome the problem, a section of the Bridgewater canal was removed and lock gates fitted at each end of the section. A pillar was then erected away from the centre of the channel. The canal section which had been removed was then mounted onto the pillar but this time both pillar and canal section had been re-designed to allow the latter to swivel, thus leaving the channel clear. Once a vessel had passed, the Bridgewater section swivelled back into place allowing liners to pass along the Manchester canal and the Bridgewater canal continue to function. It was engineering of the highest order. The total cost of the canal project, amounting to more than £15,000,000, was raised by the Manchester City Corporation and a group of local manufacturers and investors.

1855 - 1898: Burroughs, William S - American

Burroughs patented his adding machine in 1892, and his achievement was applauded by the US government, but he did not realise any financial benefit (see Blaise Pascal).

1858 - 1940: Hadfield, Richard - English

Hadfield was the metallurgist who discovered "silicon" and he developed many alloys to improve steel. His discovery of Manganese was of extreme importance, as it was the first alloy to combine great hardness with ductility.

1920 - 1995: Pilkington, Lionel B - English

Glass manufacturer Pilkington invented an ingenious method of making sheet glass, called the "Float Glass Method". It also eliminated the need to grind and polish the sheets after manufacture and the method is now used universally.

9. PHOTOGRAPHY IN FOCUS

1703 - 1771: Hall, Chester Moor - English

Hall invented the world's first refracting telescope and, in making lens from differing types of glass, discovering a combination which eliminated colour distortion, he had inadvertently invented the achromatic lens. It was well ahead of its time and later proved invaluable to those striving for a permanent photographic image.

In 1725, John Schultze a German chemist noticed that a solution containing a mixture of chalk and silver nitrate, when spread on paper, became blackened when exposed to sunlight. He was obviously interested in the subject but only academically because he did not take his experiment any further.

1766 - 1828: Wollaston, William H - English

In 1807, Wollaston invented the Camera Lucida, a four-sided prism mounted above a table on which is laid a sheet of paper. The camera is placed facing the light source and image to be copied and this is reflected by the prisms on to the paper, enabling an accurate tracing to be made.

1773 - 1805: Wedgewood, Thomas - English

Photography is about creating and retaining images from one surface to another and, in that context, it interested Thomas Wedgewood, who needed to increase production of decorated pottery. He decided that a faster image transfer could be achieved by photographic means. In 1790, by using a "Solar Microscope" of his own design, which materially intensified the light under which he laid a frond of fern on a silver nitrate sheet, by this means he created a negative. He submitted his findings to the Royal Institution in 1802 and it included comments by Humphrey Davy, who had assisted him. He achieved his goal of providing an image from which his artists could more quickly trace a picture but the negative he had produced could not be fixed.

1765 - 1833: Niepce, Joseph N - French
1789 - 1851: Daguerre, Louis J - French

Niepce followed a similar path to that of Wedgewood. In 1826, using a Camera Obscura, his subject was his courtyard, the light being reflected onto a pewter plate coated with bitumen and exposure time was about eight hours.

The result, though it was unclear and could not be reproduced, was acknowledged to be the first permanent photographic image.

After Niepce died, Daguerre persevered with the Camera-Obscura but now began projecting it onto silver plate treated with various salts.

In 1837, he announced his "Daguerreotype Process". The exposure time was reduced to 30 minutes, the pictures were clear, but still not able to be reproduced. Each one had to be made individually and enclosed in glass to prevent it being rubbed away. For a period, this method enjoyed great success.

1792 - 1871: Herschel, John - English

Herschel coined the word "photography", prior to which it was known as photogenic drawing and his contribution lay in the discovery of making "fixing salts", which he called "the Cyanotype process".

Images could be made onto glass but copies could not be taken and a blue tinge remained from the cyanide.

1800 - 1877: Fox-Talbot, William H - English

His inventions were quite independent of the experiments taking place in France. In 1835 Fox-Talbot had already registered his work two years before Daguerre. It recorded his success in making a paper negative but it had not been widely publicised and, when information came from France in 1837, it demonstrated that they had been following totally different paths. Fox-Talbot had devised different chemicals to achieve his improvements and, in 1841, he patented another new process which he named "Calotype and which produced both a positive and negative. He could also make any number of positives, which, when washed in a solution of his own device, were immune to further exposure to light. This, it was argued, was the true invention of photography. By 1851, he had developed the first "Flash Photography", by using a Leyden Jar battery to create the spark to activate the chemicals in the flash tray.

c.1805 - 1875: Petzval, Joseph - Austrian

In 1841, Petzval devised and manufactured a portrait lens which superseded all others. The larger aperture gave added illumination, which reduced exposure time, and set the standard, until the advent of the digital age.

1813 - 1857: Archer, Fred S - English

Archer gave a massive stimulus to research in 1851, by inventing the "Wet Collodium Plate", which took plate-making to another level. It was a method of converting the negatives into positives and it also enabled multiple copies to be made.

1831 - 1879: Maxwell, James C - Scottish

Maxwell, an eminent physicist in 1861 concluded that, by using three primary colours as filters, colour photographs were possible. His findings were solidly-based and, apart from minor modifications, were universally accepted.

1840 - 1900: Maddox, Richard L - English

Inventor of the Dry Plate Process in 1871, Maddox coated the plates with a gelatine solution which was impregnated with silver bromide.

The resulting sensitivity of the plates proved to be so fast, that it was possible to take a photograph with a hand-held camera. The tripod to eliminate shake was now unnecessary; the age of the amateur photographer and of modern photography was now imminent.

1843 - 1900: Abney, William - English

He evolved a number of advances, including that of gelatine emulsion and the invention of print-out-paper. His projects were developed by the Ilford Company in 1891 and put into production, demonstrating that photography could be an affordable hobby.

1845 - 1921: Lippman, Gabriel - French

The emphasis now moved towards making colour photographs. In 1908, Lippman's scientific bias directed him to using light waves instead of chemical dyes and his results were stunningly clear and accurate. Unfortunately, he also encountered the problem of being unable to make copies of the colour plates.

1855 - 1921: Friese-Greene, William - English

Arguably the inventor of cinematography Friese-Greene was an innovator who made important contributions to the medium. He provisionally patented the first camera and projector in 1899 and a system of stereoscopic photography which is the basis of 3D His last invention was a camera which could take a series of photographs on a roll, giving the impression of animation. However, it needed modification to increase the speed of the images to give accurate representation of animation and, before he had completed the necessary improvements, he was leapfrogged by Edison, who was awarded credit for the device. He had encountered one of the pitfalls of registering a patent too early.

1828 - 1914: Swan, Joseph - English

Between 1870 and 1890 Swan had become interested in Photographic processes and devised a method of producing dry photographic plates. He also patented bromide paper.

10. THE BIRTH OF RADIO

A veritable pool of knowledge was plumbed in the early days of radio and below is a list of the major contributors. Their theories and laws governing the fundamental aspects of the study were invaluable and, when combined with individual inventions, problems were approached from all angles and theory was turned into reality.

1791 - 1867: Faraday, Michael - English

In 1821, the study of the revolutions of a magnetic needle in response to an electric current provided the basis of the research which culminated in Faraday's invention of the dynamo. His research into electrolysis laid the basis of electro-chemistry. In 1837, he also showed how the charge to a condenser is stored by the non-conducting medium in electrical actions in and, in 1841, these were followed by laws governing electrical power and induction.

1797 - 1878: Henry, Joseph - American

Scientific knowledge was advanced considerably when, in 1842, Henry published his findings of the behaviour of oscillatory discharge from a condenser.

1824 - 1907: Thomson, William (aka. Lord Kelvin) - Irish

Possibly the foremost physicist of his time covering so many branches of science he had a considerable depth of knowledge in particular that of telecommunications.

1831 - 1879: Clerk-Maxwell, James - Scottish

Clerk-Maxwell was a scientist and mathematician who, in 1864, proved some of Faraday's discoveries from a mathematical basis. His own theories were expressed in his " Theory of Electromagnetic Waves"

1834 - 1913: Preece, William H - Welsh

Preece studied under Michael Faraday and, developing wireless telegraphy, he eventually became a senior figure in the Post Office. He, in turn, gave valuable advice and support to Marconi, who had come to England to further his ambitions.

1850 - 1949: Lodge, Oliver - English

Lodge invented the radio wave detector, which was fundamental to early radio/telegraph receivers. In 1897, he also patented the method of controlling the frequencies of radio transmitters and receivers, by using capacitors and conductors.

1857 - 1894: Hertz, Heinrich - German

In 1888, Hertz was the first to broadcast and receive radio waves and he proved the existence of radio waves, thereby endorsing the Maxwell Theory of Electro-magnetism of 24 years earlier. Kelvin and Faraday had also subscribed to the same theory. A "Hertz" is now used to express the frequency of alternating current and he is also credited with being the first to send and receive radio waves. Hertz observed the theory, expressed and established by Clerk-Maxwell in 1889 c; that electromagnetic waves and those of light and heat were identical.

1834 - 1937: Marconi, Guglielmo - Italian

Marconi came to England in 1896 and his practical system for wireless telegraphy was granted a patent. He first established communication between Penarth and Weston-super-Mare and, by 1898, he had sent the first two paid messages. In 1901, he transmitted Morse signals from Poldhu in Cornwall to North America, but like other scientists knew radio and light waves travelled in straight lines but they still were unable to explain what caused them to follow the curvature of the Earth.

1850 - 1918: Braun, Ferdinand K - German

Braun's research and development included the invention of special circuitry, which vastly increased the range of the sound of the Marconi transmitter and was also crucial to the clarity. In 1909, he was, alongside Marconi, awarded the Nobel prize for physics.

1834 - Michael Faraday had explained in his Laws of Electro-chemistry the relation between the amount of electricity produced and the quantity of solutions employed.

1836 - John Daniell (English) introduced the cell which bears his name and was regarded as an improvement over the Voltaic pile.

1859 - Gaston Plante (French) converted a battery into a portable re-chargeable electrical unit (often called an accumulator). This proved invaluable for householders whose property had not yet been wired for electricity and enabled them to be able to operate an early wireless set. (They were usually taken to a local cycle shop who, at a small cost re-charged them).

1866 - Georges Leclanche (French) made a brilliant adaptation of many previously known aspects, when he invented the battery in dry-cell form, which became an immediate success.

1881 - 1955: Fleming, Ambrose - English

The change from telegraphy to telephony was made possible from the moment that Ambrose Fleming invented the thermionic radio valve in 1904, leading the way to the development of actual broadcasting. Its practicality was demonstrated in 1907 and, in 1920, the first news and entertainment programmes went on air in Britain. The BBC opened its first station in 1922, using the call-sign, "2LO". Its impact on telephony was that, for the first time, it was possible to convert powerful radio signals of alternating current into weak direct current signals suitable for the telephone.

1873 - 1961: De Forest, Lee - American

In 1906, De Forest, a prolific American scientist and inventor became interested in the thermionic valve, or diode, which Fleming had invented. He improved its capacity by some modifications which provided greater amplification and named it the Audion. This was in turn replaced by the transistor, invented by Bell Laboratories USA.

1831 - 1900: Hughes, David - Anglo-American

Born in London, Hughes invented the carbon microphone in 1878, while living in America. It was the prototype of the modern equivalent and it also provided the finishing touches to the Bell telephone by the clarity it imparted to voice transmission.

1850 - 1925: Heaviside, Oliver - English (see science)

The discovery of the "Heaviside Layer" was announced in 1902 but simply as a fact, its exact location was not detected until 1925 thanks to Edward Appleton. It had obvious and far reaching implications for both Radio and Radar. (A.E Kennedy of America reached the same conclusion on its location at almost the same time.)

1895 - 1966: Eccles, Wm. Henry - English

A pioneer in radio reception based on the Heaviside theory, Eccles also contributed to the expansion of knowledge of both the wave theory and telegraphy. One of his publications, "Handbook of Wireless Telegraphy", played an important role.

1890 - 1954: Armstrong, Edwin - American

Just as Lee de Forest had improved upon Fleming's thermionic valve, Armstrong, in turn, raised the performance provided by the Audion to further refinement by inventing a new type of circuitry, the superheterodyne, which increased volume and quality. He also introduced frequency modulation FM.

In 1903, by agreement, wireless telegraphy was installed on British ships. In 1912, it proved its value in the Titanic disaster, 711 lives being saved when the system was used to summon help. By 1916, British ships were obliged, by law, to install wireless telegraphy equipment on all ships which weighed more than 3,000 tons.

11. RADAR – THE NATION'S INTRUDER ALARM

Although a number of nations were experimenting with the subject, it was the British scientists who brought radar to a successful conclusion but, because of the need for the utmost secrecy, no information was leaked to the outside world. The programme was accelerated by the conviction that the events unfolding in Europe in the early 1930s could lead to war with Germany and that Britain would undoubtedly be bombed. The need to buttress our defences was vital as, in a matter of minutes, bombers could be across the Channel, meaning that only protection in the form of technology could provide the key to instantly identifying and giving warning of their approach.

Heaviside had predicted that there was a layer in the ionosphere which deflected radio waves to conform to the curvature of Earth and , Edward Appleton, in 1925, not only confirmed it but accurately calculated its distance to be 60 miles, by transmitting a radio wave into space and noting the time the radio wave took to reach its destination and return to Earth. This clearly showed what was possible and equally clearly begged the question - could this technique be applied to any solid object to monitor its position?

Britain possessed a number of the world's top scientists and, in 1935, the government decided to pool their collective scientific knowledge to urgently provide a solution to the problem. The aim was to either to discover a ray, based on radio waves powerful enough to disable enemy aircraft, or to be able to refine the Appleton technique to make identification of enemy planes more precise.

The government already had a large and powerful radio station at Daventry, where it was investigating the possibility of locating thunderstorms by the use of radio waves. All the laboratory facilities were there and it was an ideal location to assemble the chosen team of scientists. It was decided that Robert Watson-Watt, who was the resident meteorologist at Daventry, would be appointed team leader.

Many pieces of scientific equipment already existed and their capabilities were well known to the experienced scientists within the group. One such item, the "fluorescent cathode tube", was employed almost from the start and, with it, they were rapidly able to demonstrate that the distance between the two spots or blips on the screen roughly corresponded to the distance from the aircraft being tracked. This tracking was refined by observations recorded on receivers, as an aircraft flew over at specified heights and distances. The success was rapid and by 1936 they could accurately pin-point an aircraft as far away as 75 miles. Radiolocation was now a fact and, only one year later, a chain of tracking stations was established from the Firth of Tay to the South of England

Williams, Frederick C - English

In 1939 Williams developed the first practical radar method capable of identifying friendly aircraft from the ground. He later produced an automatic identification of friendly aircraft which was fitted to fighter aircraft.

Others were directing their efforts to countering the threat to our shipping.

Once the war started, as expected, the Luftwaffe began to bomb Britain, by which time it was already possible to monitor their approach. The consistency of interception was such that German pilots soon became convinced that it could not be coincidental.

Lovell, Bernard - English

Because of his studies into cosmic rays in the 1930s, Lovell was chosen to work on radar development for detection and navigation. Although Britain had achieved the aircraft location system, we had no defence against the U-boats (submarines), which were taking a heavy toll on our shipping. It was a major individual achievement by Lovell, when he invented his HS2 device, a form of radio-television device which made submarines so easy to locate that, within six months of its introduction, monthly losses in shipping tonnage dropped by over 80%.

As the war progressed the role of radar changed from that of a defensive tool into an offensive weapon. It was adapted to bomb-aiming systems, enabling British planes to accurately bomb targets from great heights.

A new problem was presented to Britain towards the end of WWII, Germany had developed the V2, a guided missile containing high explosive material. It was fitted with a timing system calculated to cut out over London and plummet to the ground. One of the radar team members, W.S. Butement developed a radar beam, or ray, which, when directed by a defending aircraft at the missile, could override its timing device and cause it to instantly self-destruct.

Butement had emulated a similar device produced by British inventor Harry Grindell in about 1920 which generated a ray, that he actually demonstrated, could stop a motor car or a motorcycle at 50 feet, by interfering with the electrical charge to the ignition. His claim that it could be further improved to be effective against an aircraft was never taken to conclusion, partly perhaps because it was so far ahead of its time and that, sadly, he died at young age.

After the cessation of hostilities in Europe this convocation of great scientific brains, (considered by the government to be so essential to the project in 1935), was quietly dispersed to return to their former careers, if at all possible. Their collective achievement appears not to have received any general recognition (similar to that accorded to Bletchley Park) and their names for individual feats of inventiveness practically unknown.

It also seems inconceivable that this team of eminent scientists who were so closely involved in this critical and unique exercise could be so ignored. It is even more astonishing that this team of great scientists with their previous individual outstanding achievements already on record should be shouldered aside and the entire credit for the development of radar attributed (as is generally the case) only to the team leader, meteorologist Robert Watson-Watt.

Postscript

c.1860 - 1925: Hulsmeyer, Christian - German

Ironically, as early as 1904, and basing his experiments on the discoveries by Hertz, Hulsmeyer developed and patented an obstacle detector and navigation device. It apparently evoked little interest in Germany or indeed anywhere else. Perhaps the concept was too far ahead of the current technology.

12. COMPUTER SCIENTISTS

1792 - 1871: Babbage, Charles - English

Generally regarded as the Father of the computer, Babbage graduated from Cambridge University in 1814, where he occupied the Lucasian Chair of Mathematics. In 1823, he received a government grant to build his " Difference Machine", capable of calculating numerical tables but abandoned the project because he ran out of money. After his death, the details for the completion of the machine were found among his papers and were used by engineers of the Science Museum, over one hundred years later, to complete it exactly as Babbage had envisaged.

1815 - 1864: Boole, George - English

A mathematician and logician, Boole devised what became known as Boolean algebra. It was based on logical sequences, which culminated in unerring calculation. He was appointed chair of mathematics by Queen's College, Cork, in 1849, where he remained until his death. He was honoured by both Dublin and Oxford universities in 1857. The inherent accuracy of his system was realised almost 100 years later, when it proved crucial to the demands of the electronics industry in the development of computers, calculators, mobile telephones and software, all dependant on the Boolean algebra incorporated in these electronic devices.

The Enigma Project

An outstanding achievement was that of the scientists and mathematicians who broke German communication codes in WWII. Alan Turing and

Tommy Flowers were leading figures in the design and building of the computer "The Colossus", which put Britain at the forefront of world computer technology. To learn about this remarkable organisation, which is now a museum and open to the public, consult the Bletchley Park website - www.bletchleypark.org

1911 - 1977: Williams, Frederick C - English

Williams invented the Williams Cathode Ray Store Tube in 1945, which was a memory system. It was an important step at the time, giving great impetus to the industry at the beginning of the Computer age and was used universally in the first generation of computers.

b.1955: Berners-Lee, Tim - English

Inventor of the World Wide Web in 1989, Berners-Lee decided that it should be a medium of universal benefit and therefore, unusually, in the modern materialistic world, he declined the opportunity to control it for personal gain.

13. THE DEVELOPMENT OF TELEVISION

Like many other inventions, television began as a theory, followed by painstaking experimentation. Years before the radio experiments began much was already known about the basic principles governing the conversion of light waves into electrical impulses. As a consequence the names of major scientific figures occur throughout.

Faraday established the first laws of electricity and electrolysis. In about 1865, Clerk-Maxwell added to them with his electro-magnetic theory of light and waves.
Crookes' studies of the cathode had identified the source of the luminescence, resulting in his invention of the "Crookes Tube", another step toward understanding the possibilities of the cathode in the pursuit of TV development.

In England, as early as 1873, Willoughby-Smith has been named as a scientist who demonstrated how light waves could be transformed into electrical impulses. It was reported that, in 1880, an American inventor named Carey also conducted experiments with the purpose of transmitting vision over wires but without progress.

1860 - 1940: Nipkow, Paul G - German

In the 19th.century, Paul Gottlieb Nipkov produced images by using a high-speed scanning disc. Weiller, his German compatriot, achieved a similar effect using a mirror drum. These were devices to break down and then re-assemble the elements of a picture but were unable to achieve a clear image and, having decided that their systems were not the answer, discontinued further experiments.

1888 - 1946: Baird, John L - Scottish

Baird persevered with the mechanical system pioneered by Nipkow and Weiller and , on 27th January 1926 he succeeded in transmitting both a silhouette and a face but the images were still unclear indicating that it was still not a workable system. To support his experiments he was provided with research facilities by the BBC to make experimental broadcasts, first from Oxford Street and later from the new London Regional station and, in 1932, even provided with a studio at Broadcasting House. His system failed to gain acceptance by the BBC, as the picture was too grainy and woolly. Nevertheless he is referred to as the father of television' .

Electrical & Musical Industries (EMI) Hayes, Middlesex

In 1930, Alfred Clarke, a senior researcher, was appointed to organise a team to develop an alternative system to that of Baird. A number of years earlier EMI had studied the possibilities and when examining notes of previous studies, they discovered that, in 1908, Campbell Swinton, one of their physicists, had studied the cathode tube and its possibilities and opined that it offered the perfect solution to achieving picture clarity and that further work on mechanical projects were a waste of time. His notes were so precise and detailed on how a system could operate in both reception and transmission, that it was given priority development facilities. The research director was Isaac Schoenberg, a Russian electronics engineer, who had come to London and it was under his direction that a camera pick-up tube and a cathode ray tube were integrated and named EMITRON. The first stage converted the light variations into electrical impulses, which the latter then translated into a picture. This invention resulted in the world's first practical television which, in 1936, was used by the BBC to send the first public telecast from London.

14. THE TELEPHONE

1834 - 1874: Reis, Johann P - German

Using a vibrating membrane induced by an electrical current, Reis invented an early form of telephone circa 1861, however, there does not seem to be evidence that it was regarded as anything more than a teaching aid.

1847 - 1922: Bell, Alexander G - Scottish

Bell emigrated to America, where he became interested in electric sound production and teamed up with Thomas Watson, a clever electrical engineer who was able to convert Bell's ideas into a functioning piece of equipment. The result was patented in 1876, as the first voice-carrying telephone. It survived challenges against the patent and became an immediate success worldwide.

1835 - 1901: Gray, Elisha - American

Gray applied for a patent barely two hours after Bell had logged his application and, upon learning of the situation, he hotly contested Bell's patent. The salient part of his objection claimed that the transmitter which he himself had invented was eventually used in the device used by Bell but was not contained in Bell's original patent application, thereby rendering it invalid and also constituting plagiarism. The court disregarded his claim and found in favour of Bell.

1808 - 1889: Meucci, Antonio - Italian/American

An impoverished Italian immigrant to the USA who spoke little English, Antonio Meucci was involved in a complicated legal wrangle over the origin of the telephone invention and claimed that he had registered an application for a patent 16 years before that of Bell. His case was based on the equipment he had developed for use within his home, as an aid for keeping close communication with his bed-bound wife, while he was at work. Meucci had not only published details of his device as early as 1860 in a New York Italian language newspaper but followed it by a public demonstration of the instrument.

Following the granting of patent rights to Bell there ensued a long and complicated case in the US courts, in which the judges found in favour of Bell.

Italian-Americans who knew the history of the case, and all the earlier events, never relinquished their call for justice and eventually they succeeded in persuading the US government to re-examine the facts, evidence and circumstances which led to what they claimed was the usurpation of Meucci's invention and they asked that truth and historical accuracy be preserved.

On 17th June 2002, the US House of Representatives, after examining all the details and records surrounding the case, ruled that Meucci, not Bell, was the true inventor of the telephone. After a hundred and thirteen years, their claims were vindicated

1831 - 1900: Hughes, David E - English

A Londoner who lived and worked in the USA for many years, Hughes was interested in music and sound in particular. His experiments in the use of various materials resulted in his invention of the carbon microphone in 1878, which gave greater clarity in both telephone and radio transmission.

15. PROGRESS IN AGRICULTURE

Among all societies, the subject of land cultivation was important but, even until the 17th century, it was still fairly basic, with the three strip system being common, each being cultivated in rotation and one laid fallow each year. The poor returns meant that no provision of fodder was available to maintain cattle or sheep over the winter. As a consequence, the majority of animals were slaughtered in the autumn which, with a growing population, presented an increasing problem.

1591 - 1652: Weston, Richard - English

Weston went to Europe to explore the possibilities of finding grasses and other crops not common in England and might be incorporated into our farming system He discovered that the red clover and turnips were prevalent in the Netherlands and were an ideal supplement for animals when other plants were not available. The turnip was particularly valuable in winter. He introduced them to Britain and added them as a source of animal winter feed supplies.

1674 - 1730: Townsend, Charles - English

Townsend became a devotee of the turnip, lauding its valuable properties so much that he was given the nick-name Turnip Townsend. He proved its all-round worth by growing it to improve the quality of the land, after which he used it to nourish his sheep. The manure from the sheep then supplied the fertiliser for his corn. Townsend's major contribution was the introduction of the four-course crop rotational system.

1741 - 1820: Young, Arthur - English

Young was an excellent writer on agricultural practice including crop yields, rotation and nutrients and, impressed with the efficiency of the seed drill, he made some improvements to it. He studied the effect of good husbandry and the use of fertilisers on yields of wheat and recorded that, through good practice, the better farms increased their wheat crop to 30 bushels an acre, which was about a 50% increase on a decade earlier. When The Board of Agriculture was founded in 1793, he was invited to become the Secretary.

1814 - 1900: Lawes, John Bennet - English
1817 - 1901: Gilbert, Joseph H - English

These two brilliant chemists formed a partnership to try to improve farming efficiency and their discoveries proved to be of immense value. Lawes was an agriculturist, whose experiments with ammonia had provided valuable information and, in 1843, Lawes and Gilbert began to manufacture superphosphate, which led to the foundation of a new industry – that of scientifically-produced fertilisers. These new products also improved the nutrition of the animal herds, through the betterment of the grasses on which they fed.

1803 - 1873: Liebig, Justus F von - German

A chemist whose theories on soil, nutrition and the use of fertilisers proved so valuable in raising the productivity of land, that Liebig is claimed by some to have revolutionised European attitudes towards agricultural practice.

16. GENETIC ENGINEERING ENTERS THE FARMYARD

1725 - 1795: Bakewell, Richard - English

Bakewell pioneered stock improvement by examining the characteristics of various breeds. By selective cross-breeding he was able to blend the best attributes of one breed with desirable attributes of another until he had achieved an animal which when bred onward would reliably convey all of the desirable traits to its offspring. His development of the famous Leicester breed of sheep combined an animal with long coarse wool with a hardy stocky animal providing a plentiful supply of meat.

1753 - 1832: Ellman, John - English

Ellman worked along the same lines as Bakewell but selected his breeding stock to emphasise the qualities of fineness of fleece, to give higher quality yarn..

1749 - 1820: Colling, Robert - English
1751 - 1836: Colling, Charles - English

Impressed by the results achieved by the sheep farmers they followed suit by applying the same techniques to cattle and succeeded in creating the highly-regarded Shorthorn cattle of Tees-side. The production of fine pedigree stock by British farmers gathered pace, becoming a major export, with the Herefords and Shorthorns covering the ranges of the vast ranches in the American West. In Britain the Devon, Sussex and Aberdeen Angus cattle were prized for beef and the Jersey, Guernsey, Friesians and Ayrshire breeds valued for their milk productivity.

1822 - 1884: Mendel, Gregor J - Austrian

Mendel applied the same process of selectivity in his study of plants. His first experiments were with peas, the varieties of which were crossed and re-crossed, until he had established a number of strains which, from seed, exhibited the required individual characteristics - short, tall, climbing, bushy, or vigorous and productive. He published his work in 1866 but it was over thirty years before the world's leading botanists realised the importance of what has since become a major contribution to world nutrition and general plant improvement.

17. MECHANISATION ON THE FARM

1674 - 1741: Tull, Jethro - English

In 1701, Tull invented the horse-drawn seed drill, which speeded up the operation and efficiency of sowing. He later invented a horse-drawn hoe, which eliminated hand weeding of certain crops. His invention of the combined drilling, ploughing and cultivating machine was the forerunner of all modern mechanised farm implements.

He originated the modern system of farm management in Britain and, in 1731, published "Horse Hoe-ing Husbandry" which gave further impetus to agricultural improvement.

The Dutch farmers had developed a form of wooden plough, called the "shore and mould" board, which many English and Scottish farmers used. In the late 18th century, a Scottish farmer named James Small improved on this by converting it from wood to iron. In England, Robert Ransome took this a stage further, by making the first "chilled iron share" in 1779 and, a little later still, a self-sharpening share in 1803.

1719 - 1811: Meikle, Andrew - Scottish

There were several unsuccessful attempts to develop a threshing machine before Meikle invented a practical thresher. He had improved it to such a stage by 1800 that, in essence, 200 years later, today's mechanised version is still very similar, with the revolving drum and sieve system for separating the grain from the straw.

Harvesting, another labour intensive task often fell victim to inclement weather. This caused the inventors to then turn their attention to developing a practical reaper.

1799 - Englishman J.Boyce was granted a patent for a machine with rotating scythes.

1826 - Scotsman P.Bell produced a reaper which was reputedly able to work an acre an hour.

1833 - American O. Hussey patented an improved reaper

1834 - American C. McCormick patented a machine which evolved eventually to exceed all others in overall performance. It was a Reaper-binder, for use on wheat, oats or barley. It automatically formed them into sheaves and bound them with twine, even to the point of tying the knot.

18. TECHNOLOGY VERSUS DISEASE

In the 16th. century, after study and dissection of cadavers, Andreas Vesalius, an Italian physician, left details of his findings, which may have guided Ambroise Pare' of France toward the use of ligatures to control blood-loss during amputation.

The microscope in its basic form was developed in the Netherlands circa 1600. One of the claimants was Dutch physicist Hans Jansen but it was not until it came into the hands of Antonie van Leeuwenhoek, a lens-grinder, that the quality of magnification was such that it became possible to study organisms previously too small for the human eye to detect. He had unwittingly inaugurated the science of bacteriology, the importance of which was not immediately appreciated and many years elapsed before it was exploited.

In 1616, Englishman William Harvey's theory governing the circulation of the blood became central to medical doctrine. A concomitant discovery was, that through arterial injection, a different form of embalming was possible.

Chester Moor Hall invented the achromatic lens, vastly improving the microscope and its capabilities (see photography section).

William Hunter, a Scot, came to London in 1741 to practice surgery and obstetrics. He became a member of The Royal College of Surgeons and excelled in pathology and, to aid his studies, he gained government sanction to use cadavers for dissection.

His brother, John Hunter, aged 20, a carpenter by trade, followed William to act as his assistant, preparing cadavers for dissection, as well as studying anatomy and surgery. His knowledge of anatomy and physiology became equal to, if not greater than, that of his older sibling. After his death in 1793, the Government purchased his remarkable collection of specimens. Amazingly, for a man of so little formal education, his written work was said to be of the highest quality.

Edward Jenner, an English doctor, investigated smallpox and found that milkmaids commonly contracted a mild form of the disease from cows, which rendered them immune to the virulent form. He concluded that people exposed to the milder form might also develop equal immunity. In 1798, after two decades of injecting volunteers with Cowpox, he proved conclusively that his vaccination theory was effective and safe and published his findings. It was met with much trepidation in some countries but he eventually received international acclaim.

Anaesthetics had been sought since ancient times, to give relief while treatment of injuries took place. Early methods included opium or, for the sailors who were faced with battle wounds requiring amputation, copious draughts of rum. Circa 1900, cocaine obtained from the Coca plant in Peru was used as a local anaesthetic. It continued to be used in dentistry until the mid-1940s.

In 1800, Humphrey Davy, an English chemist, discovered nitrous oxide, otherwise known as laughing gas. It proved to be an effective anaesthetic but it appeared to be disregarded for surgery.

In 1818, Michael Faraday, an English scientist, discovered the anaesthetic properties of ether. It was not until 1842, when used by American doctor Crawford Long, that the first pain-free operation was recorded.

Chloroform appears to have been well-known in scientific circles but it was Scottish doctor James Simpson who, in 1847, was the first to use it as an anaesthetic in an operation.

In 1848, Louis Pasteur of France was studying fermentation of wines and made discoveries which explained the presence of organisms capable of self-reproduction. He also concluded that they were the cause of bacterial infection and putrefaction under certain conditions.

In 1865, when reflecting on Pasteur's findings on the presence of dangerous organisms, Joseph Lister, an English surgeon working in Glasgow, deduced that these organisms must be present in all hospitals, thus explaining high mortality rates especially following amputations. He insisted that all wounds, plus all areas of the operating theatre, including instruments, be treated antiseptically with carbolic acid. The recovery rate of patients, post surgery, dramatically improved.

In 1882, Robert Koch, a German physician, isolated the bacilli which caused tuberculosis and, a year later, repeated his success when he discovered the bacilli which caused cholera.

While studying a cathode-ray tube, German scientist Wilhelm Rontgen observed that the proximity of a certain chemical caused fluorescence, indicating some kind of radiation and it transpired that the radiation he had discovered could penetrate many solid objects. Further tests showed he had discovered the X-Ray, for which he received the Nobel Prize in 1901.
(Not realising the dangers from excessive radiation-in the 1920 to 1940 period, some shoe shops installed them to ensure a correct shoe fit.)

In 1897, investigating tropical diseases, Robert Ross, an English scientist, traced the life-cycle of the mosquito and discovered that its gastro-intestinal tract and its salivary glands contained the malaria parasite and that the disease was transmitted among birds by mosquito bites and therefore deduced that humans were equally vulnerable. He was assisted by consultation with Patrick Manson a Scot and Alphonse Laveran, a Frenchman, both army surgeons who had observed the parasites in dead soldiers.

In 1901, Karl Lansteiner, an Austrian, discovered differing blood groups and described three major types,A, B and C, adding A/B and RH soon after. It improved the safety of blood transfusions. He received the Nobel Prize in 1930.

In 1921, Federick Banting of Canada and Charles Best of the USA were the first to extract insulin, a pancreatic hormone which controls the release of glucose into the blood. By injecting insulin into diabetes sufferers, they were able to effectively control the disease and save many lives.

In 1942 war casualties prompted the government to urge all bacteriologists to intensify their research efforts. Alexander Fleming, a Scottish bacteriologist, examined culture plates in his laboratory and one which had been there since 1928 showed signs of resistance to bacteria which he classified as penicillin. It was in its un-purified state and the task of isolating and purifying the culture was allotted to Howard Florey, an Australian pathologist, and Boris Chain, a German-born British biochemist. Their collective achievement made medical history and all three were awarded the Nobel Prize for medicine in 1945.

Probably one of the greatest scientific achievements of modern times was made by Francis Crick, an English scientist, James Watson of the USA and Maurice Wilkins from New Zealand who, in 1962, were jointly awarded the Nobel Prize for defining the molecular structure of deoxyribonucleic acid (DNA). Crick worked in the research laboratory at Cambridge University, studying the molecular structure of living organisms in an attempt to unravel the genetic code. . Francis Crick was described as the most influential biologist of the 20th. century.

After WWII, Godfrey Hounsfield, an English physicist, led the research team of EMI company of London that designed and built the first British all-transistor computer. He later invented the CAT (computerised axial tomography) scanner for medical diagnosis, which could provide a three dimensional image of living tissue and they built the first successful machine in 1972. He was awarded the Nobel Prize in 1979.

The Nobel Committee were informed that an American physicist, Allan Cormack, had written about the possibility of such equipment but without embarking on any practical development. The Committee, however, decided that he would also share the Nobel Prize.

In 1973, research by English biologist, Ian Wilmut, resulted in the first ever birth of a calf from a frozen embryo. While remaining conscious of the moral issues, he also engaged in stem cell research, to counter degenerative disorders such as Parkinson's and Diabetes. He afterwards moved to the Roslin Institute in Edinburgh and was the head of the team which cloned "Dolly" the sheep in 1996, the first animal to be cloned from an adult cell.

19. THE SOUNDS OF MUSIC

The mouth was the likeliest source of music, followed by the drum and animal horn and the last is possibly the oldest form of wind instrument. From that basic principle came a range of instruments capable of sounds of exquisite beauty. Some stringed instruments are said to have originated in the East and, in basic form, were products of pure inventiveness and, over time, were improved by innovation. This makes it less easy to trace the origins of instruments like the lute or the guitar, which have contributed so much.

The Viola Group

The family of stringed instruments from which others evolved were made in four sizes -the viol (treble), the viola (tenor), the violincello (bass) and the violone (double bass).

From the viola group came the idea of the violin and a number of names were experimenting with various designs. In the middle of the 16th century, Gasparo da Salo, Maggini and Gaspar Tieffenbrucker of Brescia, were all engaged in studying a variety of shapes and proportions to develop the ideal instrument.

Identical research was happening in Cremona where, in the 16th.century, Andrea Amati made instruments of the viola group, before transferring his attention to the violin. It is claimed that the first instrument bearing the Amati signature is dated 1546. His sons, Antonio and Geronimo, and a grandson, Niccolo, all continued in the business. Geronimo redesigned the violincello to acquire added resonance and better tone. Circa 1700, Niccolo became independent and made violins to his own design, reducing the overall size and body shape; changes which achieved good clear tonal clarity with resonance. Before his death in 1737, the true violin began to appear.

Circa 1700, Antonio Stradivari, a former pupil of Niccolo, also began to build violins to his own design. He changed the shape, curves and arches, used different woods for various parts, scraped the body panels thinner than before and increased internal supports. He moved sound holes and developed unique varnish, (ingredients still unknown) and the result was an instrument which has retained the same construction for over 300 years.

Giuseppe Guardinieri, also of Cremona, produced instruments which can be compared to the Stradivari. His best work is dated between 1710 – 1745.

Only one factor remained absent from realising the full beauty of the Violin and that was provided by Frenchman Francois Tourte late in the 18th century when he devised the perfect bow.

Pianoforte

The possible progression of the Pianoforte was as follows:

The Monochord - later several strings were added.

The Clavichord - roughly the same instrument but with an added keyboard

The harpsichord - similar in shape to a grand piano, but with keys of different lengths, long for natural notes and short for sharps and/or flats, plucked by quills.

<u>Spinet</u> - a small harpsichord
<u>Virginals</u> -an even smaller harpsichord type which could be rested on the lap while playing.
Dulcimer, cembalo (similar) and psalter were other variations.

In 1709, Italian Bartolommeo Christofori invented the "pianoforte". Previous instruments were monotonic but his had soft and loud tones operated by pedals and the strings were struck by hammers instead of being plucked. The only major changes since then are that other piano-makers added iron frames to (a) allow increased string tension (b) enable bigger hammers to give greater sound quality and (c) eliminate warping of the frame.

The piano also embodied an ingenious device which automatically released the hammer from further contact with the string, immediately after it was struck. This has been called a "hopper" and its action allows the string to vibrate without hindrance, to produce a note of true clarity. There appears to be no information regarding its origin, so it is likely it could have been invented by Christofori and described by the English as a"Hopper", or perhaps by the Germans, whose word "Hopser" similarly means to jump or spring.

Trumpet

Known for thousands of years in many cultures.

Flute

Originated in France but redesigned by Theodore Boehm of Germany. The improvements he achieved in tone and range, by modifying the bore in the head vastly influenced the design of wind instruments.

Bassoon

Dates back to the middle ages and related to the Bass Shawm. Radically improved first by German concert artist Karl Almenrader with some modifications later by instrument maker Johann Heckel.

Clarinet

Invented by J.C. Denner of Germany in the 18th. century but was redesigned by the virtuoso players H.Klose and Buffet, a century later who developed it into the instrument we know today.

Oboe

Developed in France and used principally in fanfares (i.e. posthorn). It was transformed into the delicately-toned orchestral instrument of today, by French musicians Baret and Carte.

Saxophone

19th century invention by Adolph Sax of Belgium

Concertina

Concertina and the Mouthorgan were invented in 1829 by Charles Wheatstone.

Harp

Is so ancient that its origins are lost in antiquity. However, apparently in 1795, Sebastien Erard of

Strasbourg redesigned the European model and added pedals to control string tension. He later patented one with seven pedals, each able to produce two changes of pitch, a design now regarded as the basis of the modern harp.

20. MISCELLANEOUS

At the end of the 19th century through to the middle of the 20th, there were many ingenious devices made, most of them intended to make domestic life easier. Most were adaptations or innovations based on existing equipment invented by others but they were nevertheless labour-saving. Their names and dates have been so frequently publicised, that it was decided not to repeat them here - with one exception, the sewing machine.

This particular item had a unique place in the home. It enabled people to make or mend their own clothes and they could develop the skill to earn a living without leaving their home which was a considerable benefit for those who might be disabled.
By developing their skill they could compete on even terms with anyone.

The Sewing Machine

In 1790, Thomas Saint (English) invented the first practical sewing machine. It had automatic feed and continuous thread but it was designed for the leather trade and unsuitable for domestic use.

In 1830, Bartholemy Thimonnier (France) invented the treadle machine but the hooked needle was cumbersome.

1840 circa, Newton and Archbold (English) introduced a design using a straight needle and lock-stitching.but this machine did not attract enough interest.In 1846, Elias Howe USA) introduced a design featuring a horizontal stitching movement of the needle and a continuous feed but it lacked the complete cycle being sought.

In 1851, Isaac Singer, (US) incorporated the best features from all existing models, to which he added his own innovations, resulting in an outstanding machine. He then made a unique offer that, for those without the wherewithal to pay for it outright, people could purchase a machine on instalment credit terms. This meant that it could be paid for while, at the same time, using it to earn a living. Unsurprisingly, he rapidly became the world's biggest manufacturer of sewing machines and an immensely rich man. With every invention, there seems always to be a lurking innovator who can make improvements and Singer's brilliant machine was no exception to the rule.

In 1854, Alan B. Winslow (English) added a four-motion-feed and a metal-toothed plate to carry the cloth forward and then return to repeat the motion.

As always, because of the "law of unintended consequences", there are occasions when an action, decision or innovation might provide additional benefit or incite a worse situation. Two such actions are demonstrated here.

Food preservation has been a perennial concern to mankind. Salting or smoking were some of the methods used successfully on a relatively small scale. The need achieve the ability to provide food on a grand and continuous scale became crucial to Napoleon Buonaparte when his French armies were invading almost every other country in Europe. To find a solution the French government offered a lucrative prize. The winner was Nicolas Francois

Appert, who in 1810 invented a method of preserving food by canning, little realising that his action would not only bestow enormous benefit to households in the future but assist the continuing carnage in Europe in the short term.

In 1882, James Dewar, a Scottish chemist, devised a method of storing liquefied gases at required temperatures. It consisted of an inner vessel separated from an outer by a vacuum, which reduces loss or gain in temperature of the contents. This device proved to have unexpected practical benefits on a domestic level, when it became the "thermos flask" used in almost every family picnic.

Notice of Liability

Every effort has been made to ensure that this book provides accurate information and the author shall not be held liable for any loss suffered as a result of the use of any information contained within.

Questions & Answers

A. Questions on The Railway

Who is regarded as the Father of the locomotive?
Richard Trevithick

Can you name the inventor who in 1801 was the first to carry passengers by using steam power?
Trevithick, with his road vehicle

Where was Trevithick's (and the world's) first working steam locomotive installed: England; Scotland; Ireland or Wales?
Wales - to haul coal

Who was the inventor and founder of the railways and where was he born?
George Stephenson, Newcastle, England

George Stephenson's first engine had trials in 1814. It was hauling 30 tons – what speed did it reach 4, 10, or 16 miles per hour?
4 mph

Where and in what year did Stephenson install the world's first public railway?
Stockton-on-Tees. September 1825

The world's first modern railway was installed in 1830. Was it in Newcastle to Middlesborough or Manchester to Liverpool?
Manchester to Liverpool

In 1813 Which inventor improved by five-fold, the weight-hauling ability of Trevithick's Locomotive and how was it achieved?

John Blenkinsop. In 1813 he invented the rack-rail system

Can you name of the locomotive inventor who introduced the use of high-pressure steam for locomotion?
Trevithick

How did Trevithick overcome the accumulation of water in Cornwall's deep mines?
He designed and built a high-pressure steam engine

Can you name any significant works designed and built by Robert Stephenson?
The Menai railway bridge linking Anglesey to mainland Wales; the high-level bridge at Newcastle-on-Tyne; the Victoria bridge over the St. Lawrence in Canada

From 1813 One man's patented locomotive the famous "Puffing Billy" worked continuously until 1865 when it was taken to the Kensington Museum. Can you name him?
William Hedley

At what date were the first railway sleeping cars introduced between England and Scotland 1860 – 1873 or 1881?
1873

Who in 162 was granted a licence to operate a private bus service in Paris?
Blaise Pascal

How did George Stephenson who lacked a formal education manage to master the technical knowledge to develop the locomotive?

He spent every week-end dismantling and re-assembling the Killingworth colliery engines until he mastered the method of construction

In what year did George Bradshaw publish his famous railway time-table?
1839

What was the date of the first railway excursion organised by Thomas Cook?
1838

The great liner the Queen Elizabeth was built on the Clyde – Who built the engines?
Charles Parsons of Newcastle-on-Tyne - the inventor of the Steam Turbine

B. The Automobile : A Personal Stagecoach

In 1876 Nikolaus Otto produced the first practical, and successful, four-stroke internal combustion engine - what prevented its continued production?
Beau de Rochas in 1862 had patented a similar theory in France but though he never attempted in any way to prove his theory, Otto's rights to manufacture were over-ruled

The use of rubber for the manufacture of many products, including tyres and raincoats owed their existence to which English chemist?
Thomas Hancock in 1821 invented the Masticator to convert raw latex into rubber

Who invented hot vulcanisation of rubber and when and what was his nationality?
Thomas Goodyear - 1844 - American

Not quite a motor car but what was the invention of Gottlieb Daimler of Germany in 1883?
The motorcycle

Daimler surprised everyone with his second invention – any idea what it was?
The motor boat

One great German engineer, invited to England in 1913, to discuss his engine with the British Admiralty but disappeared from the German ship bringing him over – who was he?
Rudolf Diesel

From the workshops of Frederick Bosch in Germany came two refinements to the internal combustion

engine which improved its performance enormously -
can you name them?
The Magneto and the Spark plug.

Who was regarded as the Father of the British Motor
Industry who imported the first I.C.E. into England?
Frederick Simms of London

Two motor manufacturers in Britain in the 1920's
competing for the mass-market produced cars for
£100, can you name either?
Morris Motors and Ford

One company made the world's first turbo-jet car
which reached a speed of 150mph plus. Can you name
it? And the date?
The British Rover Co. Jet-1 - 1952

Which company registered the first patent to
manufacture inflatable rubber tyres in1888 - (a)
Goodyear (b) Pirelli (c) Dunlop?
Dunlop

When was the London to Brighton Run established?
1896

What was the reason behind the establishment of the
London-to-Brighton Run?
*To celebrate the increase of the speed limit from 4 to 14 mph
and abolishing of the flagman*

What Government action brought about the increase
of the speed limit in 1896?
*The passing of the "Light (Road)Locomotives Act of 1896"
in Britain*

C. Pedal Power - Goodbye Shanks's Pony

Who was the first to take out a patent for a pneumatic tyre?
Robert Thomson in London in 1845. They were for carriages and made of leather

Who had the primary claim to be the first to add traction to the early bicycle?
Kirkpatrick McMillan a Scottish blacksmith in 1839 when he fitted a shuttling device to propelled the machine forward.

Who made the first machine to be classed as a bicycle an what date?
Pierre Lallement of Paris in 1866. In England called "Boneshaker."

Where and when is it claimed the first cycling club was founded?
The Pickwick Club of London. In 1870

Which bicycle design was regarded as the first true bicycle and by which maker circa 1870?
The first to be described as a true bicycle was the "Ariel" built by Englishman John K. Starley

What qualified the 'Ariel' bicycle to be so described?
It had centrally mounted pivotal steering, equal- sized wheels plus his invention of tangentially-spoked wheels providing tensile strength

Who is generally described as the Father of the bicycle industry?

John Starley

When was the chain added to the bicycle?
In 1874. It is attributed to H.J.Lawson an English engineer

When were solid tyres changed to pneumatic?
In 1888 a Scottish vet living in Dublin used the "hot vulcanisation" method to register a patent of the first rubber inflatable tyre

D. Spinning & Weaving

Who was the first recorded inventor to attempt to mechanise stocking making?
William Lee in the 16th century

Who invented the "Flying shuttle" and when?
John Kay's patent in 1733 quadrupled the speed of the loom and founded automatic weaving

Two men invented the first power spinning machine in 1738 not superseded until 20 years later by Arkwright- who were they?
Lewis Paul & John Wyatt

In 1748 Paul patented a carding machine for the weaving industry-what did it do?
It removed all foreign substances and laid the fibres in one direction

In the 18th century Jaques Vaucanson of France visualised a method of controlling the loom but failed in practice. About 200 years later another Frenchman succeeded, What was his name?
Joseph Jacquard

What method was used in the Jaquard / Vaucanson system?
Punched cards

James Hargreaves invented the Spinning Jenny in 1764 What change did he make?
He mounted them vertically in multiples.

In which year did Richard Arkwright invent his famous "Arkwright Frame"?
1769

How did Samuel Crompton improve on Hargreaves/Arkwright inventions to develop his "Mule"?
He simply combined them before inventing the spindle carriage to reduce breakage.

How did Jedediah Strutt become wealthy in the 18th century?
He resurrected William Lee's 16th century invention of the stocking frame and perfected it

Edmund Cartwright invented the power loom for weaving in 1785 - why was it surprising?
Because he was a country vicar without previous experience of the textile industry!

Another inventor used the punched card method in his work – who was he?
Charles Babbage in his computer.

Can you name the inventor who improved the Crompton model by adding the "self acting Mule" circa 1840?
Richard Roberts

About 1850 Samuel Masham devised a wool combing machine –what was different to existing machines?
It almost eliminated waste

William Cockerill and his brother installed the latest manufacturing techniques into Belgium. Can you describe what they installed?

They built spinning machines and carding machines

Which English émigré to New England, in 1793, founded the American cotton spinning industry?
Samuel Slater

E. The Age of Steam

Which nobleman in 1645 invented a steam pump capable of raising water from a depth of 40 feet?
The Marquis of Worcester

What method for raising steam pressure was first advocated by Papin?
The piston and cylinder

Who in 1698 made the first practical steam pump?
Thomas Savery

In 1715 one man used a piston to compress steam efficiently and laid principles for later modifications - who was he?
Thomas Newcomen

When and by whom was the next attempt made to improve steam power for industry?
1759 John Roebuck began to financially support James Watt to study the Newcomen engine

Which Birmingham manufacturer supported James Watt after Roebuck went bankrupt?
Mathew Boulton

In which year was Watts improved engine patented?
1768

In 1781 Jonathon Hornblower invented the first reciprocating compound engine which out - performed all others – why did he not market it?
Boulton & Watt claimed infringement of patent. Legal action was too costly for Hornblower

In 1804 Arthur Woolf invented a compound high pressure engine developing almost twice the output of the Watt model. Its final patent was withheld until 1810 – Why?
To avoid legal conflict with Boulton & Watt whose patent rights terminated then

In 1800 Trevithick developed a high-pressure engine for the Cornish tin miners – why was it necessary?
Current machines were low pressure and lacked power to cope with ever deeper mines and water accumulation. He did not violate any patents – Watt did not make H.P. machines

Did anyone ever build a roadworthy vehicle driven by steam?
Yes , Goldsworthy Gurney built a passenger carrying vehicle capable of 15 m.p.h. and for a short time attempted to operate a service between Bath and Gloucester

Who was the inventor of the steam turbine which powered the great Queen liners in the 1930's?
Charles Parsons of Newcastle-on-Tyne

F. How Aviation Got Off The Ground

Who is regarded by many historians as the true inventor of flight?
George Cayley

What was the basis of his theories?
Contours of wing and body contours. In short, Aerodynamics

Who was the glider designer who ,under practical conditions, truly established and proved the principles of Aero-dynamics?
Otto Lillienthal

What distance did the Wright Bros achieve in their first flight (a) 1800 ft.- (b) 852 ft.- (c) 475 ft.?
852 ft.

Who was the first in Britain to build and fly his own plane?
Alliot Roe

In 1909 Alliot Roe made a circuit of London in a tri-plane what was the engine size: (a) 25 hp. (b) 19 hp or (c) 9hp?
9 hp. J.A P. engine

Who won the final Schneider Trophy race in 1931. Name the plane, designer and speed?
Britain in the Supermarine designed by R.J. Mitchell speed 407.5 mph

A short list of firsts, each of which can form an additional question:

1903 Wright Bros achieved first ever powered flight

1908 Alliot Roe. Achieved flight with his first plane of (75 feet)

1909 Louis Bleriot crossed the Channel in his monoplane (powered by an Anzani engine 25 hp)

1910 Charles Rolls – first to fly non-stop across the Channel and back.

1910 A.V.Roe began manufacturing planes-AV504 trainer/bomber sold thousands –WW1

1913 Nesterov & Pegoud inaugurated art of aerobatics by flying a loop-the-loop

1919 John Alcock & Arthur Brown (pilot / navigator) fly the Atlantic

1926 Alan J. Cobham flew to Cape Town and back and repeated the feat to Australia

1927 Charles Lindbergh flew solo from New York to Paris. Time 35.5 hours.

1930 Amy Johnson-first woman to fly solo from England to Australia – Time 20 days.

1931 Francis Chichester- first to fly the Tasman sea east/west to Australia

1935 Amelia Earhart flew from New York to Ireland - solo

G. Chronological List Of Science & Scientists

Which scientist is regarded as the originator of experimental science?
Roger Bacon

For which controversial belief(at the time) is Copernicus particularly remembered?
That the Sun was the centre of the Universe around which all the Planets revolved.

Can you name the calculating device devised by John Napier?
Logarithms

In the early part of the 17th century the inventor of the first astronomical telescope was enabling scientific assertions to be made. (flouting religious teaching) - who was he?
Galileo

Can you name the mathematician / astronomer who described the laws governing the planetary system?
Kepler

Name the scientist who proved that electricity can be generated by friction. An early electric generator. Who was he?
Guericke

After the Great Fire of London who was responsible for the reconstruction of the City?
Robert Hooke

Who rebuilt St. Paul's Cathedral and also built 52 churches in London?
Christopher Wren

Apart from his well-known work on gravitation what practical instrument for the study of astronomical bodies did Isaac Newton invent?
The reflecting telescope

In 1682 Edmund Halley predicted (correctly) that the comet named after him would return-in which year, 1759 -1789 - 1799?
1759

Can you name any other important work carried out by Halley?
He catalogued all the stars and constellations of the southern hemisphere during a two year stay on the island of St. Helena in the 17th century

Two scientists were recognised by the Royal Society as simultaneously inventing the quadrant in 1730, one was John Hadley of England can you name the other one?
Thomas Godfrey. American

Can you describe the original purpose for which the quadrant was intended?
The calculation of the altitude of stars

For what other practical purposes did the quadrant proved valuable?
Navigation, gunnery ,surveying

John Hadley improved on the quadrant when he invented another instrument what was it?

The sextant

In his invention of the thermometer what changes did Fahrenheit make in the material used to register the temperature on the scale?
His first gauge used alcohol and the second mercury

What was discovered from the Leyden Jar experiment?
Musschenbroek and Kleist had independently demonstrated that electricity could be stored.

What significant relationship does the Leyden jar have with modern science?
It has undergone much development and as a capacitor is essential to many electronic devices.

Anders Celsius designed a thermometer – How did it differ from the Fahrenheit system?
It used a different calibration 0 for freezing and 100 degrees for boiling point of water.

Joseph Priestley was famous for his discovery of Oxygen – but what did insist on calling it?
Dephlogisticated Air

Priestley also discovered a method of obtaining a number of gases can you name any of them?
Hydrogen, hydrochloric acid, nitrous oxide, ammonia, carbonic oxide, sulphur dioxide, nitric acid

In general terms what subject does Coulomb's Law cover?
Electromagnetism

Who invented the first battery and what was its purpose?
Alessandro Volta. To deliver continuous current

What instrument for the measurement of liquid density did William Nicholson invent?
The Hydrometer

Can you name the discoverer of the element Tungsten or Wolfram?
Wilhelm Scheele

Tungsten was first isolated by Fausto and Juan Elhuyar - for what is it most used?
To increase the hardness of steel and filaments for light bulbs.

Who discovered Palladium and Rhodium?
Wm. Wollaston

Which scientist developed "The Atomic Theory of Matter " and is regarded as one of the founders of "Physical Science"?
John Dalton

Whose wave light theory gave a calculated measurement of seven colours named by Newton
Thomas Young

A scientist who is considered to have founded the science of electrodynamics and also invented A machine to measure electric current- The Ammeter. Can you name him?
Andre Ampere

He eliminated- colour distortion- by his invention of the "Achromatic Lens". His name?
Peter Barlow

Who is reputed to be the first to prove the relationship between electricity and magnetism?
Hans Oersted

He discovered sodium, potassium, chloride and "Laughing gas – Who was he?
Humphrey Davy

He produced the first practical electro-magnet and invented the –Commutator.?
William Sturgeon

He gave his name to the Law describing the theory of Voltaic current. Name the Law?
Ohms Law

The scientist who supported and expanded Young's theories of wave-light phenomena And revolutionised lighthouse emissions in the process. Can you name him?
Augustin Fresnel

John Daniell invented the Hygrometer- what was its purpose?
To measure humidity

Who is often referred to as the Father of Electricity?
Michael Faraday

Before Samuel Morse devised his famous code, he was awarded a prestigious medal in another profession. What was it?

Painting. He received a Gold Medal from the Royal Academy.

What were the names in the partnership which invented the electric telegraph?
Wheatstone and Cooke

Can you list any other inventions by Wheatstone?
Concertina – electric clock - stereoscope

Why did Charles Darwin delay publication of his findings for so many years?
Because of the risk of offending religious bodies

The co-operation between Prof. Plucker scientist and Heinrich Geissler a Glassblower saw the invention of the Geissler Tube – What was its significance?
Cathode and Vacuum Tube development is at the base of much of electronic science.

What discovery by Robert Bunsen benefitted astronomers?
That each element emitted light with an individual wavelength

Who introduced the first Law of Thermo-dynamics?
James Joule

Which scientist when seeking a cure for hydrophobia followed Jenner's solution to reduce dramatically the effect of a communicable disease?
Louis Pasteur when he used vaccination to counteract Hydrophobia.

Which famous Doctor followed the researches by Pasteur, to dramatically reduce the post- operational death rate in hospitals?
Joseph Lister

He invented the incandescent lamp and the miner's electric safety lamp –who was he?
Joseph Swan

Mathematician who wrote the theory of light and revolutionised electrical theory?
James Clerk-Maxwell

He discovered Thallium, in 1879 and later identified the polarity of particles emanating From the Cathode and the source of the luminescence. Can you name him?
Crookes Dark Space. William Crookes

Who catalogued the "known" elements into seven segments called The Law of Octaves"?
John Newlands.

Who re-classified the increasing number of elements In his work "The Periodic Law"?
Dimitry Mendeleev

John Rayleigh and William Ramsay discovered and isolated Argon and Neon in1894. Which others did they add four years later?
Krypton and Xenon

Who discovered Helium in the atmosphere?
Joseph Lockyer & Pierre Janssen, observed it independently during the same eclipse in 1868

Who discovered Helium on Earth?
William Ramsay in 1895

Who discovered the location of a dependable radio wave reflective layer in the in the ionosphere?
Edward Appleton in 1947

Who invented a radio wave detector?
Oliver Lodge

What did J.J. Thomson name the particles that he discovered within the atom?
He called them Electrons

Who was the originator of the Quantum Theory?
Planck Max

Why did Marie Curie name the radium active substance that she discovered-Polonium?
In honour of Poland where she was born

Whose researches into radio-activity assisted the " Curies" in the joint acquisition of the Nobel Prize?
Henri Becquerel

His study of nuclear reaction and conclusions regarding future progress brought him world renown and expanded the importance of radio-activity within physics. The accuracy of his predictions were verified in 1932 by Cockcroft & Walton - do you recognise the scientist by the description?
Ernest Rutherford

Marconi received his patent in England and from there he made the world's first radio transmission.

Give the date and names of the locations engaged in this radio link-up.
Poldhu Cornwall & St. Johns, Newfoundland. In 1901

They built the first nuclear particle accelerator named the Cockcroft / Walton generator which enabled them to announce that they had split the atom. Can you name the year?
1932

When first Russia and then the USA launched their first satellites into space they then realised that they lacked the ability to track their travel. Which scientist was able to provide the information and from where. Can you name him and the location of his radio telescope?
Bernard Lovell from Jodrell Bank, Manchester

H. Engineering Inventions & Innovation

Archimedes proved his famous principle in a famous demonstration- what was it?
He proved, by his " Principle" that the crown made for King Hiero II was not pure gold

Could someone explain how it was proved that Hiero's crown was an alloy?
Because it was lighter than one of pure gold it displaced a lesser amount of water when immersed

Who was the first to use coke (instead of charcoal) in the furnaces for smelting iron?
Dudley Dud was granted a patent in 1621.

The first calculating machine was invented in 1647. Can you name the inventor?
Pascal Blaise

Abraham Darby also used coke –How did his method differ from Dud's?
He used larger furnaces generated greater heat and produced finer metal

In 1742 he invented "Sheffield Plate" by fusing silver on to copper. Who was he?
Thomas Boulsover

Who revolutionised the construction of lighthouses using dovetailed blocks and hydrated lime?
John Smeaton

Which versatile engineer linked the North Sea and Atlantic by building the Forth/Clyde canal?

John Smeaton

What was regarded as Robert Stephenson's crowning achievements in bridge building?
The Great St. Laurence River Bridge in Canada

Who improved upon Newcomen's steam engine adding impetus to the Industrial Revolution?
James Watt

Where was Watt's steam engine produced?
At the Birmingham factory of Mathew Boulton who also financed the project

Was Watt's engine designed to operate under high or low pressure steam?
Low

He patented machinery for "puddling" and rolling iron, at the time a major advance-Name him?
Henty Cort in 1783

Invented- Hydrostatic press, thief-proof locks, a Bank-note printing machine. Who was he?
Joseph Bramah

Who patented, in 1789, the first lifeboat capable of negotiating rough sea conditions?
Henry Greathead

Thomas Telford was a prolific builder of canals. Can you name any of his major work?
Caledonian Canal - The first Menai bridge.

Scheele discovered - Wolfram /Tungsten - but who isolated it and used it to harden steel?

Fausto and Jose Elhuyar

Who invented the Tunnelling shield- essential to sub-aqueous tunnelling?
Marc Isambard Brunel

Give the date and project in which the Tunnelling Shield was first used?
1825 = Rotherhithe to Wapping Tunnel

Who was in charge of the Rotherhithe-Wapping tunnel?
Isambard Kingdom Brunel

Name the inventor whose method of making pulley blocks were 10 times faster than before?
Marc Isambard Brunel

In 1783 he invented a machine to increase iron production by "Puddling" His name please?
Henry Cort

Who invented the Hydrostatic Press, thief-proof locks and Banknote printing machines?
Joseph Bramah

In 1799 he patented the first Lifeboat capable of coping extreme sea conditions - Who is he?
Henry Greathead

Two brothers isolated Tungsten –added as an alloy it makes the finest tool-steel, their
Spanish – Fausto & Juan Elhuyar

Arguably the founder of machine tool-making and a major figure in the Industrial revolution?

Henry Maudsley

He founded the British rubber industry by his machine to convert Latex into rubber –His name?
Thomas Hancock - who invented the "Masticator" in 1821

Portland cement patented by Joseph Aspden in 1824 included what special ingredient?
Silicon

What did Joseph Whitworth introduce in 1851 that was of great benefit to engineers?
A precision made range of nuts and bolts of varying sizes – off the shelf

Which English émigré was regarded as the Father of the American Flint Glass industry?
Benjamin Bakewell circa 1810.

Who invented the lawnmower?
Edwin Budding 1830

Who designed and built the Clifton Suspension Bridge?
Isambard K. Brunel

For what was Wm. Armstrong of Newcastle famous?
He invented the hydraulic crane, and the hydro-electric machine (delivering frictional electricity).

Name the inventor of pre-stressed concrete.
Eugene Freyssinet 1938

He invented a "safety hoist" for goods-into what did it evolve and what was his name?

In became The elevator (or in Britain - Lift) by Elisha G. Otis

Which invention helped Brunel to reduce shipbuilding construction time?
James Nasmyth with his invention of the steam hammer

This man invented the method of mass-producing steel?
Henry Bessemer

Name the inventor of Plastic and the name he gave to it?
Alexander Parkes and he named it Celluloid

What other product did he invent?
Parkersine a watertight adhesive for the Mackintosh coats

Two English engineers in partnership built the undernoted - Who were they?
- The Metropolitan underground railway, London
- Victoria station - Pimlico bridge and numerous railway systems-circa 1860
- The Aswan Dam (Assouan) in Egypt for hydro-electric power 1898-1902
- The Forth Bridge 1889

Benjamin Baker & John Fowler

Alexander Eiffel is famous for his tower but what other great work is attributed to him?
The Panama Canal in which he was among the first to employ compressed-air caissons.

Invented in 1892 his adding machine, ironically he had no personal need for it, Why?

He was applauded by the US government but it failed commercially

A metallurgist who discovered Silicon, more importantly Manganese, the first alloy to combine great hardness with ductility. Who was he?
Richard Hadfield

Which Lancashire glassmaker invented the" Float Glass" method to mass- produce glass sheet?
Lionel B. Pilkinton

I. Photography

Who discovered that images could be created by exposing silver salts to light?
Joseph Schulze

What had Schultze demonstrated by his experiment with silver salts 1725
Some of the basic optical and chemical principles of photography

Which potter invented a method to fix images on pottery or glass?
Thomas Wedgewood

When inventing his "transfer of image to glass" method, what medium did Wedgewood use?
Silver salts to create the image and a light box of his own design to project it

Who made the first permanent photograph?
Niepce 1826

When did Daguerre present his photographic process?
January 1839

When did Fox-Talbot patent the first paper negative from which any number of positives could be made?
February 1839

Which invention of Frederick Scott-Archer improved the method of making negatives?
The "Wet Collodium Process" which used glass plates

What accessory did Petzval make so well in 1841 it was not superseded until the digital age?
The portrait lens

Which mathematician concluded that colour photographs were possible using three primary colours as filters?
Clerk-Maxwell

In 1871 Richard Maddox invented the Dry Plate Process why was a tripod no longer essential?
The sensitivity of the plate was so quick it eliminated shake when used as a hand-held camera

Which London company developed Wm. Abney's inventions of Print- out- Paper and gelatine emulsion?
The Ilford Company in 1891

Who patented the first camera and projector in 1899 and is credited with the invention of Cinematography?
William Friese-Green

J. The Birth of Radio

Which inventor described the behaviour of the oscillatory discharge from a condenser?
Joseph Henry in 1842

He was a senior figure in developing wireless telegraphy for the Post office -his name?
William Preece

Who invented the radio wave detector fundamental to radio/telegraph receivers?
Oliver Lodge

He proved their existence and was first to broadcast and receive radio waves in 1888?
Heinrich Hertz

He was the first to patent a practical wireless telegraphy system in 1896?
Guglielmo Marconi

He invented special circuitry increasing the range and clarity of Marconi's transmitter and was the joint recipient of the Nobel prize in 1909-Do you know him?
Ferdinand Braun

Who discovered the reason why radio waves did not fly off into space?
Heaviside when he discovered an ionised layer in the upper atmosphere causing deflection

Who located the Heaviside layer and how?

Edward Appleton. He sent a radio signal into space and timed the response

A "Handbook of Wireless Telegraphy" was published by a pioneer in radio who was he?
Henry Eccles.

The change from telegraphy to telephony was made possible by the thermionic valve making broadcasting possible –Who invented it?
Ambrose Fleming in 1904

What was the call sign of the first radio station?
2 LO

Two years after Fleming's invention of the thermionic valve it was improved by whom?
Lee de Forest whose development was named "The Audion"

Do you know what replaced the Audion?
The transistor invented by Bell Laboratories

This scientist introduced a cell which improved the original Voltaic Pile?
John Daniell

He converted the battery into a rechargeable unit known as an "accumulator" essential for early sets when many houses were without direct electricity – Have you heard of him?
Gaston Plante

In 1866 he brilliantly adapted many well-known aspects of electrical knowledge to produce a battery in the form of a dry-cell?
Georges Leclanche

Another scientist who responded to various innovations by developing a new type of circuitry to increase volume and quality of sound "The superheterodyne"- what is his name?
Edwin Armstrong

Who invented or developed Frequency modulation FM?
Edwin Armstrong

K. Radar - The Nation's Intruder Alarm

How did Appleton determine the distance of the Heaviside layer?
He transmitted a radio wave into space and timed the outward & return journey

What might be deduced from the Appleton experiment?
That this method might be expected to work when dealing with solid bodies

What did the government do in 1935 to investigate the possibilities of aircraft interception?
They assembled a group of top scientists to pool their brains and experience

Where was the chosen team of scientists concentrating on Radar located?
Daventry radio station

How soon was the Radar team equipped to reliably track an aircraft?
Within a matter of months up to 15 miles and 75 miles after a little more than a year

How soon was radar available to the R.A.F.?
At the beginning of the war they were fully trained in the use of radar

What other uses other than identification were made of radar?
It was adapted to bombers as an aiming device giving almost unerring accuracy. It also became an extremely effective U-boat tracking device

One of the early aims was to develop a powerful Ray
a weapon was it ever produced?
*Yes. It was an effective weapon against the V2 doodlebug
and caused them to self-destruct.*

Was there ever any precedent to the Ray weapon?
*Yes, in the 1920's Harry Grindell made a similarly effective
but more basic device*

L. Computer Science

Who is regarded as the Father of the computer?
Charles Babbage

What was his invention named?
The Difference Machine

It is said that Babbage did not complete his machine how then was its efficiency validated?
After his death the drawings and papers for its completion were found and It was successfully completed by engineers from the Science Museum

What was one of the key components in the design?
A punched card system

Who first invented the idea of control by punched card system?
Vaucanson

Which English mathematician devised a system of algebra perfect for computer technology?
George Boole invented "Boolean" algebra

Which team of scientists placed Britain at the head of Computer technology in World War 2?
The code – breakers of Bletchley Park

In 1945 he converted a known scientific device (the Cathode Tube into a memory system crucial to the first generation of computers – Who knows his name?
Frederick Williams

In which year did Tim Berners – Lee invent the World Wide Web?

1989

M. The Development of Television

The origins of television are more complicated than the versions which we have all come to accept and do not devolve on one man. In an attempt to clarify matters questions have been set to match the progress.

He passed beams of intense light through tiny holes onto a white background and firmly established the wave nature of light even to being able to almost exactly measure the wavelengths of colours. Name the scientist and in which decade of the 19th century?
Thomas Young (between 1800-10)

Who discovered photo-conductivity of selenium leading to invention of photo-electric cells used in early television sets. He also proved that light waves could be transformed into electrical impulses?
English scientist Willoughby-Smith in 1873

Who were the first to achieve picture images?
Nipkow & Weiller produced images by using large spinning discs but the quality was unclear

Were there any other attempts made?
Yes, Vladimir Zvorykin filed for a patent in the USA in 1925 but no more was heard of it

When did John Logie Baird apply for a patent?
1926

What system did he develop?
He used the one pioneered by Nipkow & Weiller but with more success

Why was his system rejected by the BBC?
It was basically a silhouette and a face and an image. The picture was too woolly and grainy and unsuitable for broadcasting

Who produced the workable TV system eventually used to broadcast?
Electrical & Musical Industries (EMI) Industries Hayes, Middlesex

Was it based on the Nipkow/Weiller-Baird method?
No That was a mechanical method EMI took an electronic route

So who was responsible for the EMI successful television equipment?
It was a team effort, In 1930 Alfred Clarke a senior researcher was appointed team leader

Were EMI forced to start from scratch?
No They found notes from a study of 1908 when another physicist had stated categorically a Cathode tube was the perfect answer for both transmission and reception

Was that the only change to the earlier study notes?
No. A member of the team Russian émigré Isaac Schoenberg changed the format to a camera pick-up tube and a cathode ray tube which were integrated

What were the functions of the pick-up tube and the cathode ray tube developed by EMI?
The former converted the light variations into electrical impulses the latter translated them into a picture. and in 1936 the world's first practical television broadcast was made by the BBC

N. The Telephone

Who is reputed to have developed an early basic telephonic device?
Johan Reis circa 1860 during an experiment into sound reproduction

Who applied for the first patent of a workable telephonic device?
Antonio Meucci in 1860

On what date did Alexander G. Bell register his telephone patent?
1876

Were there any other challenges to Bell's patent?
Elisha Gray applied for a patent only two hours after Bell

On what grounds did Gray protest against Bell's patent?
He maintained that the transmitter used was plagiarised from his design and was not part of the original patent presented by Bell. His challenge was over-ruled

Why was the Meucci patent disregarded?
It had elapsed and he could not afford to renew it

Did Meucci contest Bell's right to the invention?
Yes, with the assistance of his supporters

Who was eventually credited with inventing the telephone following examining the case by the US House of Representatives in 2002?
Antonio Meucci

O. Progress in Agriculture

Who in the 16th century introduced foreign crops to improve farm animal nutrition?
Robert Weston introduced The Turnip and Red Clover to England.

What other methods were instrumental in improving farming methods and by whom?
Charles "Turnip" Townsend introduced the four crop rotational programme and used the turnip to great effect (a) as a root crop to improve the land (b) nourish and improve sheep numbers and (c) benefit from the fertiliser from the extra sheep numbers to grow his corn.

Who advocated the adoption various techniques to improve farming practice?
Arthur Young in the 18th century wrote and advised on land management and the use of fertilisers. In 1793 he became Secretary for Agriculture at its inauguration.

What did Young prove by his insistence on good farming practice and recording results?
That by good husbandry and the correct use of chemical fertilisers wheat production grew 50%

Who produced the first chemical fertilisers and when?
John Bennet Lawes founded the first agricultural research station in the world where artificial fertilisers were manufactured. His factory opened in 1843.

On what evidence did Lawes base his introduction of these products?
He and his partner Joseph Henry Gilbert had over a long period studied plant nutrition and its effect on the health of

grazing animals. The results were overwhelmingly favourable.

What were the products first demonstrated?
Lawes first discovered a method of releasing phosphate from natural sources and patented the process

Which 19th century German chemist also advocated inorganic chemistry in agriculture?
Justus von Liebig

P. Genetic Engineering in the Farmyard

Robert Bakewell (circa 1750) was a pioneer in which stock improvement technique?
Selective breeding. His first success was the " Leicester" sheep breed which produced a fine quality wool together with a stocky meaty frame.

Was selective breeding done in the 18th century by others?
John Ellman used Bakewell's method to breed animals giving fine and high quality wool.

Can you name any who applied this method to cattle?
Robert and Charles Colling for example (mid- 18th c) produced –The Shorthorn breed. NB. (Charles Darwin published his work in 1839.which also influenced thinking)

A Monk studied plant cross- fertilisation to improve productivity and quality -Who was he?
Gregor Mendel in 1866

What was discovered by Mendel to be the secret of cross fertilisation?
Similar to Bakewell he discovered how the hereditary qualities were transmitted by noting their distribution in each successive generation.

Q. Mechanisation On The Farm

Who was the originator modern farming systems in Britain?
Jethrow Tull. His book Horse-Hoeing Husbandry is said to be the basis of later improvements

Can you name the first of Jethrow Tulls's inventions?
The "horse-drawn" seed drill and the "horse-drawn doe" circa 1700

What was Jethrow Tull's most important contribution to farming?
The invention of the first combined drilling, ploughing and cultivating machine.

Which country developed the wooden "shore and mould board" for ploughing?
Holland

Which farmer improved the plough - and what was the change?
James Small who had it made from " cast iron"

Who made further improvements to the plough?
Foundry owner Robert Ransome later made the "chilled Iron share "

Who made the first practical threshing machine in 1800?
Andrew Meikle

Who developed the first practical reaper in 1800?
James Boyce

Name American engineers who developed mechanised farm equipment?

Obed Hussey who made improvements on reapers.

Cyrus McCormack who sought perfection finally inventing a reaper- binder that not only cuts wheat, oats and barley – forms them into sheaves and even ties and knots the binding twine.

Who first marketed the Combined Harvester – and when?

Massey Harris in 1941

R. Technology Versus Disease

In which century did Ambroise Pare an innovative surgeon first use ligatures, during amputation, to restrict loss of blood?
16th

A basic version of the microscope was invented by whom and when?
By Hans Jansen about 1600 He was head of a small team also credited with the idea.

Who, by improving the lens quality of the microscope, inadvertently opened the door to the science of bacteriology?
Antonie van Leeuwenhoek

In 1616 William Harvey 's theory became central to medical practice – what was it?
The circulation of the blood

He invented the "achromatic lens" apart from – photography - what did it benefit?
It made a vast improvement to the microscope for medical purposes.

In the 18th century the government bought the notes and specimens left by an outstanding Pathologist. Can you name him?
John Hunter

Who was the 18th century doctor who invented the use of vaccination to cure Smallpox?
Edward Jenner

Anaesthetics were sought for obvious reasons but it was some time some time before doctors felt confident enough to use them. In 1818 Michael Faraday discovered the anaesthetic qualities of Ether. Who was the first to use it in the world's first recorded pain-free operation?
Crawford Long in 1842.

Who proved that Chloroform, a known gas, could be used safely as an anaesthetic and when?
Surgeon James Simpson in 1847

Who In 1845 when he discovered organisms capable of causing bacterial infection could be claimed to be the world's first bacteriologist?
Louis Pasteur

Louis Pasteur cured another scourge acquired from dogs?
A vaccination for rabies

How did Joseph Lister reduce post-operative deaths so dramatically?
He insisted on all open wounds and areas of treatment be sterilised with carbolic acid

A scientist who discovered the bacilli which caused tuberculosis and a year later that of Cholera?
A= Robert Koch made his discoveries between 1881-1883

Who invented the X-Ray machine and when?
William Rontgen, 1895

Without realising the dangers which body of retailers installed X-Ray machines and why?

Shoe shops. People used to take their children to see if the shoes they were a proper fit.

He traced the life-cycle of the malaria parasite within the gastro-intestinal tract of the mosquito Who was the scientist?
Ronald Ross

What was significant about the discovery by Ronald Ross?
It opened up to development large areas of the world previously uninhabitable.

Karl Lansteiner made a discovery, which made safer, the transfusion of blood?
He identified five different blood groups

Three men accomplished an astounding breakthrough in medical science by developing an anti-biotic medicine which has saved many thousands of lives. Who were they?
Alexander Fleming, Howard Florey and Boris Chain when they discovered Penicillin

They declared they had uncovered the "Molecule of Life" in 1953 who were they?
Francis Crick, James Watson and Maurice Wilkins (for defining the DNA molecule)

Who was described in scientific circles as the most influential scientist of the 20th century?
Francis Crick

Which physicist invented the CAT scanner for medical diagnosis?
Geoffrey Hounsfield in 1972

He led research resulting in the cloning of a sheep in
1972
Ian Wilmut

S. The Sounds of Music

In 1709 the Pianoforte as invented by Bartolommeo Christofori – How did it differ from previous types?
It had loud and soft pedals and the strings were struck by hammers, not plucked

During which period were the violins made?
The middle of the 16th century

With which town are the greatest names of violin making associated?
Cremona

Whose name is most associated with the concept and design of the classic violin?
Andreas Amati

What instruments did Andreas Amati make before transferring to the violin?
The larger models of the viola family

Are you able to name the instruments of the Viola group?
The Viol – Viola –The Bass - Double Bass

Which of Andrea Amati's grandsons produced what is regarded as the first true violin?
Niccolo

Stradivarius and Guardinieri are celebrated as the finest ever violin makers- who trained them?
Niccolo, the grandson of Andreas Amati.

How and by whose genius was it possible to enhance a Stradivari?
Francois Tourte : He created the perfect bow to release both its power and delicacy.

Who invented the modern piano?
Bartolommeo Christofori in 1709

What important changes did Christofori make in his design of the piano?
Fitted soft and loud pedals plus hammer action. It was now, no longer monotonic!

Have there been any other modifications to the piano?
German makers fitted iron frames to eliminate warping of frames and increase string tension.

Who while re-designing the Flute is said to have influenced other makers of wind instruments?
Theodore Boehm

Which German concert player completely modernised the Bassoon?
Karl Almenrader

Which instrument was known in the middle ages as the Bass Shawm?
The Bassoon

The Haut Bois (high wood) was re-designed by Rudolph and Baret Carte- by what name is it now known?
The Oboe

Who invented the Saxophone?
Adolphe Saxe

What have the trumpet, clarinet and saxophone in common?

They all feature strongly in jazz music as well as classical